BASIC reliability engineering analysis

Butterworths BASIC Series includes the following titles:

BASIC aerodynamics
BASIC artificial intelligence
BASIC business analysis and operations research
BASIC business systems simulation
BASIC differential equations
BASIC economics
BASIC electrotechnology
BASIC fluid mechanics
BASIC forecasting techniques
BASIC hydraulics
BASIC hydrodynamics
BASIC hydrology
BASIC interactive graphics
BASIC investment appraisal
BASIC material studies
BASIC matrix methods
BASIC mechanical vibrations
BASIC molecular spectroscopy
BASIC numerical mathematics
BASIC operational amplifiers
BASIC reliability engineering analysis
BASIC soil mechanics
BASIC statistics
BASIC stress analysis
BASIC surveying
BASIC theory of structures
BASIC thermodynamics and heat transfer

BASIC reliability
engineering analysis

R. D. Leitch BSc, PhD
Lecturer in Statistics, Operations Research and Mathematics,
Royal Military College of Science, Shrivenham

Butterworths
London   Boston   Singapore   Sydney   Toronto   Wellington

All rights reserved. No part of this publication may be reproduced or transmitted in any form or by any means, including photocopying and recording, without the written permission of the copyright holder, application for which should be addressed to the Publishers, or in accordance with the provisions of the Copyright Act 1956 (as amended), or under the terms of any licence permitting limited copying issued by the Copyright Licensing Agency, 7 Ridgemount Street, London WC1E 7AE, England. Such written permission must also be obtained before any part of this publication is stored in a retrieval system of any nature.

Any person who does any unauthorized act in relation to this publication may be liable to criminal prosecution and civil claims for damages.

This book is sold subject to the Standard Conditions of Sale of Net Books and may not be re-sold in the UK below the net price given by the Publishers in their current price list.

First published 1988

© **Butterworth & Co. (Publishers) Ltd, 1988**

British Library Cataloguing in Publication Data

Leitch, R.D.
  BASIC reliability engineering analysis.
  1. Reliability engineering. Applications
  of computer systems. Programming languages.
  Basic language
  I. Title
  620′.00452′02855133

  ISBN 0-408-01830-5

Library of Congress Cataloging-in-Publication Data

Leitch, R. D.
  BASIC reliability engineering analysis.

  Includes bibliographies and index.
  1. Reliability (Engineering)—Data processing.
  2. BASIC (Computer program language)   I. Title.
  TA169.L45   1988   620′.00452′02855133   88-10574
  ISBN 0-408-01830-5

Typeset by Mid-County Press, 2a Merivale Road, London SW15 2NW
Printed and bound by Hartnolls Ltd., Bodmin, Cornwall

Preface

When I first conceived the idea of a book such as this, I believed it could be a pocket manual for every engineer or manager with a particular interest in reliability, as well as a text for engineering undergraduates. Now that it is complete I think that it meets the requirements of both. The book describes reliability activities as they occur during an industrial development cycle; these of necessity will include statistical analysis, although the theory has been kept to a minimum, and the more complex analysis has been restricted to Chapter 6. Concepts have been introduced in the order that they would occur in a typical industrial development cycle. Thus a project manager who wishes to use the book as a reference will progress through the book as his project itself progresses.

Chapter 1 briefly introduces the BASIC that is used in the programs in the book. Chapter 2 is introductory, and discusses the way reliability is considered in different parts of the development cycle. Chapter 3 introduces the basic concepts of reliability as a function of time, failure rate, and some basic statistical concepts. Chapter 4 deals with the modelling of complex systems and related topics such as availability and maintainability, and the idea of specifying reliability in the same way that other performance characteristics are specified. Chapter 5 describes the activities that can go on early in the development cycle, when it is necessary to analyse and possibly predict reliability. Chapter 6 is the statistics chapter, and gives some of the techniques that can be used to analyse data that is generated during development or later in the cycle when equipment is in use. This chapter describes reliability growth and Weibull analysis, and compares classical analysis techniques with Bayesian analysis. Chapter 7 is a brief look at quality assurance, offering the reader an acquaintance with the concepts involved, using inspection by attributes as a vehicle to introduce the ideas.

There are, of course, topics that have had to be omitted. Stress-strength, or interference, modelling, quality control and Taguchi techniques for example. The book could only be so long, and I apologize in advance for leaving out somebody's pet topic. The topics vi Preface

chosen were those most likely to be met by an engineer or manager during a project.

I wish to thank those who helped me during the production of the MS for the book. Martin Sandford gave some useful comments on parts of the text, and Simon Miller assisted by developing some of the programs. I typed the manuscript myself, and so the usual appreciation to a secretary must be missed, but Peter Smith and Mike Iremonger, the editors, deserve all the thanks I can give them for their encouragement to me while writing the book, and for their patience and persistence during a long wait.

Contents

Preface   v

1   Introduction to BASIC

1.1	Elements of BASIC	1
1.2	Elementary operations and output	1
1.3	Inputting data	3
1.4	Comments	3
1.5	The IF THEN and GOTO statements	4
1.6	Loops	5
1.7	Variable types	6
1.8	Arrays	7
1.9	Data statements	9
1.10	GoSUBs	11
1.11	Further reading	11

PROBLEMS   12

PROGRAMS

1.1	Quadratics	2
1.2	Mean (average) of a data set	5
1.3	Mean and variance	9
1.4	Volume and surface of a sphere	9
1.5	Density of an alloy	10
1.6	Sum of x to the kth power	11

2   Introduction

2.1	The definition of reliability	13
2.2	The achievement of reliability	14
2.3	References	17

viii Contents

3 Reliability as a function of time

3.1	Probability	19
3.2	Component and non-repairable systems reliability	20
3.3	The bathtub curve	23
3.4	Models for failure rate	25

WORKED EXAMPLES 29

PROBLEMS 36

PROGRAMS

3.1	Calculating estimates of reliability and failure rates	25
3.2	Reliability at a given time or time to reach a given reliability	28

4 Systems modelling

4.1	Reliability block diagrams	38
4.2	Maintainability, availability and replacement	42
4.3	Degraded reliability	47
4.4	The specification of reliability	54

WORKED EXAMPLES 55

PROBLEMS 64

PROGRAMS

4.1	Probability of running out of spares	46
4.2	Number of spares required	46
4.3	Solving equations using the Gaussian elimination	51

5 Predicting reliability during the design stage

5.1	Parts count and parts stress	67
5.2	Failure mode effect and criticality analysis	68
5.3	Criticality matrices	76
5.4	Fault tree analysis	78
5.5	Minimum cut sets	80

PROBLEMS 85

6 Estimating reliability

6.1	Reliability growth	87
6.2	Uncertainty in results	94

6.3	Classical statistical analysis	94
6.4	Probability distributions	95
6.5	Confidence intervals	102
6.6	Bayesian estimation	106
6.7	Further distributions	109
6.8	Weibull analysis	126

WORKED EXAMPLES 131

PROBLEMS 137

PROGRAMS

6.1	Reliability growth analysis using the Duane model	93
6.2	Significance tests	99
6.3	Confidence limits	104
6.4	Simple Bayesian data analysis performed on binomial, Poisson or experimental data	118
6.5	Weibull analysis	130

7 Quality assurance

7.1	Introduction	141
7.2	Multiple sampling schemes	144

PROGRAMS

7.1	Simple sampling scheme for a given LQL and UQL	147
7.2	Analysing sampling plans	149

Index	155

Chapter 1

Introduction to BASIC

1.1 Elements of BASIC

A computer is a machine for manipulating data, which is coded in strings of zeros and ones (bits). The computer can perform manipulations such as the elementary arithmetic operations of addition, subtraction, multiplication and division. By the appropriate use of these operations the computer can perform other more complex operations such as extracting square roots and taking logs, etc.

To be able to carry out the appropriate manipulations in the correct order, the computer must be given a set of instructions, called a program. These instructions have to be written in a well defined manner, using a restricted choice of symbols, and a set of rules specifying the way that the symbols can be used. This assemblage, of symbols and rules, is called a programming language.

The language BASIC (Beginner's All purpose Symbolic Instruction Code) which was used to write the programs in this book, was developed at Dartmouth College, USA as an easy to learn, general purpose language. It is not only easy to learn but also easy to use, particularly for simple programs found here. It is not the author's purpose that this should be an instruction manual BASIC. However, this chapter is an introduction to the BASIC that is used here.

1.2 Elementary operations and output

Suppose we wish to write a BASIC program to solve the quadratic

$$2x^2 + 5x - 3 = 0$$

using the formula

$$x = \frac{-b \pm \sqrt{(b^2 + ac)}}{2a}$$

for the solution of the quadratic

$$ax^2 + bx + c = 0.$$

2   Introduction to BASIC

A suitable program could be:

```
10  A=2
20  B=5
30  C= -3
40  X1=( -B+SQRT(B*B-4*A*C))/(2*A)
50  X2=( -B-SQRT(B*B-4*A*C))/(2*A)
60  PRINT X1,X2
```

Program 1.1 Quadratic equations

When the program is run, then the solution is shown on the screen:

0.5   -3

We must now take this program apart.

The program is divided into six numbered lines, with a single command on each line. The exact numbering is not important; just the order, to ensure the commands are carried out in the correct order. As for the commands themselves, they are of three types in this case. The first three lines tell the computer what values to give to A, B and C. We say it assigns values to the variables A, B and C. The next two lines actually perform the calculation. The * is the symbol used as a multiplication sign by the computer, the / is the division sign, and the + and − are the usual plus and minus sign. SQRT means take the square root, and the brackets after SQRT tell the computer what it must take the square root of. The results of the calculation are assigned to the variables $X1$ and $X2$. The reader will notice that there are other brackets in the expressions on these lines. This is to ensure that the arithmetic operations are carried out in the correct order. The last line tells the computer the output which is written to the visual display unit (VDU) screen.

There are simple ways Program 1.1 can be improved.The square root has to be worked out twice, and although in this example it is not so critical, in a larger, more complex program it would be good practice to add another line (line 35 maybe), and alter lines 40 and 50, so:

```
10  A=2
20  B=5
30  C= -5
35  R=SQRT(B*B-4*A*C)
40  X1=( -B+R)/(2*A)
50  X2=( -B-R)/(2*A)
60  PRINT X1,X2
```

Comments 3

1.3 Inputting data

It would also be useful if this program could be used to solve other quadratics. It would be possible to alter the first three lines every time a new problem was produced, but this would be an uneconomic use of time. Instead an INPUT command is used. Delete lines 10 to 30 and instead put in the line

20 INPUT A,B,C

If this is now run, the screen shows a question mark. It is waiting for the operator to types in the values of A, B and C, in the correct order. It can be tested by putting in the previous values, and the screen will look something like:

? 2,5, − 3
0.5   − 3

The values of A, B and C are put in in the correct order, each separated by a comma from its neighbour. The results are output in the usual way.

1.4 Comments

There is still room for improvement, even in a program as simple as this. There is the very important point of giving comments. A stranger coming across the program, or even its programmer after a long period, may have difficulty in working out what the program is supposed to do, and what the inputs and outputs are. Comments can be put in the program itself by the use of the REM command. Line 10 illustrates this:

10 REM   This program solves quadratics

REM tells the computer to ignore the rest of the line, and so the programmer can put in explanatory remarks to aid his memory, and other people's understanding of his work.

The instance when comments are of value are when the program is being run, so the operator (who may not be the programmer) knows what inputs are expected, in what order, and what the outputs are. Alter lines 20 and 60 to read:

20 INPUT 'Input the coefficients of the quadratic − x
squared first',A,B,C

60 PRINT 'The solution is',X1,X2

4 Introduction to BASIC

On running the program, the comments between the quotation marks (' ') are output to the screen, the the operator sees:

Input the coefficients of the quadratic – x squared
first? 2,5, – 3
The solution is 0.5   – 3

1.5 The IF THEN and GOTO statements

The second point is that not all quadratics have a real solution. For example, the quadratic

$$x^2 + 2x + 3 = 0$$

cannot be solved in real terms, as

$$b^2 - 4ac = -8$$

and a negative number does not have a real square root. If an attempt is made to solve this quadratic with the program as it stands, the computer will tell the operator in no uncertain terms that it cannot extract the square root of a negative quantity, and will stop running the program. It would be wise to test $b^2 - 4ac$ to see if it is negative, and if this is the case, stop the program gently with a more polite message than the computer would give. This is done by means of an IF statement, and the altered version of Program 1.1 is shown below:

```
10   REM this program solves quadratics
20   INPUT Input the coefficients of the quadratic – x
     squared first',A,B,C
30   R = B*B – 4*A*C
40   IF R<0 THEN GOTO 100
50   R = SQRT(R)
60   X1 = ( – B + R)/(2*A)
70   X2 = ( – B – R)/(2*A)
80   PRINT 'The solution is',X1,X2
90   STOP
100  PRINT 'The quadratic does not have real roots'
```

There are three points that the program now illustrates. First, line 40 has the IF statement. This is a command of the form IF something THEN command. The something is a comparison of two expressions – in this case two simple expressions, and the comparison is either true or false. (R is either negative or it is not.) If the comparison is true, the command after the THEN is carried out, otherwise it is ignored, and the next line is considered. Any BASIC command can be put after the THEN, and in this case the command is GOTO 100,

Loops 5

which is an instruction to the computer to go to line 100 of the program, and carry on from there. In this case line 100 tells the operator the quadratic cannot be solved, and the program stops. Line 40 is called a branch point of the program, as it may jump from line 40 to another point in the program. A GOTO command can appear at any point in a program, not only after an IF statement, and it can command the computer to go to any line, with a lower or higher line number in the program, as long as the line numbered appears with a command.

The second point is the way that R now appears in the program. It is first defined as '$b^2 - 4ac$', and then, if this is not negative, line 50 defines R in terms of itself. The computer is quite happy with this command, it will take R, extract the square root, and then redefine R as this new value. So if, for example, R started off as 9, on taking the square root 3 is obtained, and R is redefined as 3. This brings out the point that the equals sign used in the BASIC commands is not equality as usually defined in arithmetic and algebra, meaning 'is the same as', but rather means 'becomes'. BASIC statements are not equations, but are assignments.

The third point is line 90. The STOP command is obvious – the program will cease running. If line 90 were not in the program, then after solving the quadratic, and outputting the solution, the operator would be told that the quadratic did not have real roots, as this would be the next command in the program.

1.6 Loops

Now consider the following problem. A program to calculate the mean (average) of a data set is required. The input will be the data, and if necessary the number of data items in the set. The following program will do:

```
10  INPUT 'How many data',N
20  SUM = 0
30  FOR I = 1 TO N
40  INPUT 'Data',X
50  SUM = SUM + X
60  NEXT I
70  AV = SUM/N
80  PRINT 'The mean is',AV
```

Program 1.2 Mean (average) of a data set

This program must be taken apart much as the previous one was. Lines 10 and 20 need no explanation. Lines 30 and 60 go together,

6 Introduction to BASIC

and form what is called a FOR NEXT loop, or just a loop. It is necessary to perform all the commands between lines 30 and 60 a number of times (depending on the number of data items), and the loop tells the computer to perform these commands N times, where N is input as the number of data. I is the counting variable, and tells the computer how many times it has gone round the loop; I starts off as 1 (line 30), is increased by 1 at line 60, and if it has reached the value of N, the computer goes on to the next command of line 70. Otherwise it returns to the start of the loop, at line 40. I can appear in the loop itself if required, but it must never be redefined, i.e. there must never be a statement in the loop of the form 'I = something'. It is also bad practice to have a GOTO that jumps out of or into a loop.

Line 40 needs no more explanation, line 50 redefines SUM each time the loop is gone round, by adding X to it, the value of the data being entered, and so the value of SUM is the sum of all the data when the loop is exited. Line 70 calculates the average, by dividing SUM by the number of data, and line 80 outputs the result.

1.7 Variable types

An understanding of the concept of different variable types is not essential to be able to write technical BASIC, but it has been used in some of the programs in the book, and as it is not a difficult concept, it was thought to be worth mentioning here.

Suppose we wish to write a program that can manipulate letters, words and strings of symbols. We might wish to input 'yes' or 'no' when, for example, a program asks if we wish to analyse more data or stop the program. Using Program 1.2, we shall alter the final version so that instead of just stopping, it asks the operator if he has more data he wishes to analyse. If the response is Y (for yes) it will go back to the beginning again. If the response is N (for no) it will stop running. The following lines must be added to the program:

```
 90  INPUT 'Do you wish to analyse more data',AN$
100  IF AN$ = 'Y' THEN GOTO 10
110  IF AN$ = 'N' THEN GOTO 160
120  REM AN$ does not equal either Y or N. Give the
     operator
130  REM another chance as he may have made a mistake
     with his input.
140  PRINT 'The response is either Y(yes) or N (no). Try
     again!'
150  GOTO 90
160  END
```

Arrays 7

The required input on line 90 is not a number, but a letter. To indicate to the computer that the variable will not be given a numerical value, but what is called a string (of symbols), it is written with a $ sign after it, as AN$. Notice that string variables can be compared in an IF THEN statement, like numerical ones. However, the Y and N that AN$ is compared with must appear in quotes. In a program (though not when they are typed in on the screen as input) string variables are always written with quotes.

If the amended program is run, and it is required to find the mean of the four numbers 1, 2, 3 and 4, and then the mean of the five numbers 11, 12, 13, 14, 15 the screen should look like this:

```
How many data items? 4
? 1
? 2
? 3
? 4
The mean is      2.5
Do you wish to analyse more data? Y
How many data items? 5
? 11
? 12
? 13
? 14
? 15
The mean is      13
Do you wish to analyse more data? n
The response is either Y (yes) or N (no). Try again.
Do you wish to analyse more data? N
```

Note that the computer knows the difference between upper and lower case letters!

1.8 Arrays

Suppose we wish to write a program that calculates not only the mean of a data set but its variance. If $x_1, x_2, \ldots, x_n$ comprise the data, then the mean is defined as

$$\bar{x} = \frac{\sum x_i}{n}$$

and the variance as

$$s^2 = \frac{\sum (x_i - \bar{x})^2}{n}$$

8   Introduction to BASIC

The mean can be calculated as it was in the previous program, but to calculate the variance, we need to know the mean. And by the time it is calculated, the values of the data are lost to the program.

The data is given as a list of numbers, with subscripts to identify each element of the list. x_1 is the first data, x_2 the second, and so on. BASIC writes its subscripts in brackets, so that X(1) is the first element of the list, X(2) the second, etc. The loop that collects the data would look like this:

```
40  FOR I = 1 TO N
50  INPUT 'Data',X(I)
60  SUM = SUM + X(I)
70  NEXT I
```

This looks just like lines 30 to 60 of Program 1.2, except that X has been replaced by X(I). This is not surprising, as it does much the same task, namely it inputs the data and adds it all together. The only difference is that in the Program 1.2 the data is lost by the end of the loop, while in this program it is saved in the variables X(1), X(2), ..., X(N). This list of subscripted variables is called an array, and in this case is called the array X. The computer is very happy to deal with arrays, but it must know in advance how big the array is going to be (or how many items there will be in the list). This is done with a DIM statement, and for the above program we might add a line:

```
10  DIM  X(20)
```

indicating that X is an array with no more than 20 elements (no more than 20 items in the list). Some versions of basic allow arrays to be dimensioned dynamically, and the first two lines could be:

```
10  INPUT 'How many data items',N
20  DIM X(N)
```

Array variables can be manipulated just like any others, and the finished program could be:

```
10   DIM X(20)
20   INPUT 'How many data items',N
30   SUM = 0
40   FOR I = 1 TO N
50   INPUT 'Data',X(I)
60   SUM = SUM + X(I)
70   NEXT I
80   AV = SUM/N
90   SUM = 0
```

Data statements   9

```
100  FOR I = 1 TO N
110  SUM = SUM + (X(I) − AV)*(X(I) − AV)
120  NEXT I
130  VAR = SUM/N
140  PRINT 'Mean and Variance',AV,VAR
```

Program 1.3 Mean and variance

Contrary to good practice, this program has no comments. Readers should examine it until they are satisfied that they understand what each line does and why it is there.

The arrays described above are called one-dimensional, and can be thought of as representing lists. There are times when it is necessary to have two-dimensional arrays, with two subscripts, that can be thought of as representing tables, so that $X(3,4)$ can represent the element in the third row and fourth column of a table X. These arrays have to be dimensional as well, and a typical dimension statement in a BASIC program could be

```
20  DIM X(10),Y(20),T(20,20),S(15,12)
```

1.9 Data statements

Data statements are a useful way of assigning values to variables that are going to remain fixed throughout the program. For example, suppose we wish to write a program that calculates the volume and surface area of a sphere, where the radius is the input. π is always the same, and rather than put in a line

```
10  PI = 3.14159
```

it could be written

```
10  DATA 3.14159
20  READ PI
30  INPUT R
40  A = 4*PI*R*R
50  V = 4*PI*R*R*R/3
60  PRINT 'Volume is,',V,'Area is',A
```

Program 1.4 Volume and surface area of a sphere

Line 10 is ignored by the computer until it comes across line 20. On finding the command READ, the computer finds the line with the DATA command, and assigns the value there to the variable PI. This is a somewhat trivial example, but the following example is a little less so, and shows the saving in programming time that can be made.

10   Introduction to BASIC

An alloy is made up of five constituents, c_1 to c_5, with densities (specific gravities) of 10.1, 9.6, 12.3, 15.5 and 13.2. A program is required to calculate the density of an alloy made up of these five constituents in any proportion. The result is the weighted mean of the five constituent densities, where the weights are the proportions of the individual constituents. The following program will perform the calculation.

```
10   REM This program calculates densities
20   DIM D(5)
30   FOR I = 1 TO 5
40   READ D(I)
50   NEXT I
60   READ AN$
70   REM Input the quantities of each constituent
80   DIM Q(5)
90   TOT = 0
100  PRINT 'Input the amount of each constituent in a
     typical sample'
110  FOR I = 1 TO 5
120  INPUT Q(I)
130  TOT = TOT + Q(I)
140  NEXT I
150  SUM = 0
160  REM Calculate the weighted sum
170  FOR I = 1 TO 5
180  SUM = SUM + Q(I)*D(I)
190  NEXT I
200  SUM = SUM/TOT
210  PRINT AN$,SUM
220  DATA 10.1,9.6,12.3,15.5,18.2,'The density of your
     sample is:'
```

Program 1.5 Density of an alloy

There are a number of points to note in this example. First, the data statement has been put at the end of the program. The DATA statement can go anywhere in the program and the computer will find it. Second, there are six values assigned to variables. The computer runs down the list of values in the statement, assigning the first value to the first variable, the second value to the second variable, and so on. The programmer must ensure that the values in the DATA statement are in the correct order, and correspond to the variables he want to assign them to. Third, we can assign any type of value to the appropriate variable (assuming it is correctly defined), and in this example have assigned a value to a string variable.

Further reading   11

1.10   GOSUBs

In many programs there is a process (or processes) that we may wish to repeat a number of times in different parts of the program. Rather than write a chunk of program out a number of times, we tell the computer to go to another line, rather like the GOTO command, but in this case the computer will go back to the original place in the program when it has finished the process with a RETURN command.

For example, suppose we wish to calculate the value of

$$x_1^k + x_2^k + \cdots + x_n^k$$

for values of $x_1, x_2, \ldots, x_n$ and k that will be inputs into the program. The following is a suitable program, with a GOSUB that calculates the kth power.

```
10   INPUT 'How many data items',N
20   DIM D(N)
30   PRINT 'Input the data'
40   FOR I=1 TO N
50   INPUT D(I)
60   NEXT I
70   INPUT 'The value of K',K
80   REM Now do the calculation
90   SUM=0
100  FOR I=1 TO N
110  P=D(I)
120  GOSUB 200
130  SUM=SUM+PK
140  NEXT I
150  SUM=SUM**(1/K)
160  PRINT 'The value is',SUM
170  STOP
200  PK=1
210  FOR J=1 TO K
220  PK=PK*P
230  NEXT J
240  RETURN
```

Program 1.6 Sum of x to the kth power

1.11   Further reading

There are at least as many variants of BASIC as there are machines that run the language; for that reason it is advisable for any

12 Introduction to BASIC

programmer to read the appropriate manual. Otherwise the following are useful.

Kemeny, J. G. and Kurtk, T. E. *BASIC Programming* Wiley (1968)
Monro, D. M. *Interacting Computing with BASIC* Edward Arnold (1974)
Alcock, D. *Illustrating BASIC* Cambridge University Press (1977)

PROBLEMS

(1.1) Write a program to evaluate the expression

$$\ln\left(\frac{\sin \omega nt}{(n^2 - \omega^2)}\right)$$

(1.2) Improve your program that evaluates the expression

$$\ln\left(\frac{\sin \omega nt}{(n^2 - \omega^2)}\right)$$

paying particular attention to the values of those parameters that make the denominator of the fraction zero or the argument of the ln function zero.

(1.3) Write a program that will input N and K, both integers, and calculate

$$1^K + 2^K + 3^K + \cdots + N^K$$

Note: You can calculate the Kth power of a number in two ways when K is an integer. $X**K$ or $X\char`^K$, depending on your computer, returns X to the power K. Or if you are feeling confident, you can write a loop to do the calculation. This last technique involves two loops in the finished program, one inside the other. These are called nested loops.

(1.4) Write a program that asks for the operator's name, and after it is input, say 'Hello, ...', giving a personal touch. Note that a string variable can have a value that consists of several symbols in a line – somebody's name, for example.

(1.5) Write a program that inputs N data $x_1, x_2, \ldots, x_n$, and outputs the first k moments, where the rth moment is defined as

$$x_r = \frac{\sum (x_i - \bar{x})^r}{n}$$

$$\bar{x} = \frac{\sum x_i}{n}$$

Chapter 2

Introduction

Reliability, or more often unreliability, affects us all. We must have all discussed at some time or another the reliability of a piece of equipment, whether a complex system such as a car or an unsophisticated device like a tin opener. Reliability is primarily an engineering subject, although other disciplines, such as statistics and good management, are needed to produce reliable equipment, whether in its development, manufacture or use. Further expertise, such as psychology and ergonomics, will also be needed to produce the most reliable equipment.

This chapter discusses the definition of reliability, the measurement, or estimation, of reliability (essentially a statistical problem), and the achievement of reliability, and the problems that may arise in considering them.

2.1 The definition of reliability

The reliability of a piece of equipment or a device is its ability to perform a specified task, for a specified time, in a specified environment. It is measured as a probability. Consider these three characteristics in more detail.

'Specified task' is often called the item's function or mission. It may be easy to state the function, but for many pieces of equipment the definitions of function and failure need careful consideration. For a simple item, such as a light bulb, the definitions of function and failure are easy to define. But even for something as simple as a tin opener, failure can be difficult to define. An old, worn opener may still open tins, if used with care or by the right person. On a less trivial note, if one of the headlights on a car has failed, does the car still function? It will still carry passengers from A to B — if it is only to be used during the day. When considering a simple component there are usually two ways in which it can fail: through drift, which could be due to wear, fatigue, or some other ageing process, or catastrophically, when it breaks. The two are not necessarily unrelated. In the former case the tolerance limits must be defined,

13

14 Introduction

within which the component is still considered to be functioning. In either case we shall consider a component or system to be in one of two possible states — it is either functioning satisfactorily, or it has failed. Degraded reliability will not be considered, except for a brief discussion on Markov models.

The second characteristic is 'specified time', or possibly some other measure of ageing stress. It is usual to think of age in terms of time, but statements such as 'This car has done 70 000 miles, and has never let me down' are often heard. The point of this simple example is that age may be measured in time, distance travelled, operational cycles, etc., or a combination of two or more such stresses.

'Specified environment', or the external stress, must be stated when discussing achieved or target reliability. This will include extremes of temperature and humidity, dust and the chemical environment, as well as things that are more difficult to quantify, such as operator training, and the amount of handling the item is likely to experience. A piece of equipment cannot be expected to perform well in an environment that is outside the design envelope.

2.2 The achievement of reliability

To achieve equipment reliability, reliability must be considered as important a parameter in the equipment specification as weight, speed, cost, fuel consumption etc. These other parameters are considered long before the piece of equipment is built or even designed. To understand the way in which equipment is produced from the original idea it is essential to have some understanding of the activities involved in producing the finished product. These activities are shown in diagrammatic form in Figure 2.1, and are called the development cycle.

This book is written around a typical industrial development cycle. Before a more detailed examination of this cycle, and the activities therein, it must be pointed out that not all development cycles (possibly under another name) will exactly follow this pattern. However, the cycles will be broadly similar, and can be broken down into just five broad phases. These are specification, design, development, manufacture and use.

The first phase is the specification phase. The problem of how we should specify reliability will be examined in more detail in a later chapter. For the time being we shall say that the reliability, or more realistically, the availability, must be considered as a user requirement, with the same weight as other operational requirements, such as power consumption, speed of operation and weight. The specification may arise through an interative process, no

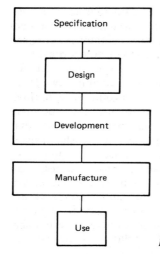

Figure 2.1 A typical development cycle

doubt accompanied by a number of paper exercises, studies of previous models and those of the competitors (if any), discussion with potential users and other market research.

Next comes the design phase. The designer must be able to justify his design, and it is at this stage that any models are refined, and some prediction work is done. Parts stress can be used to compare alternative designs; failure modes effects analysis and fault tree analysis are useful for examining the reliability structure of a complex system.

During development, when prototypes are produced and tested, data becomes available which when analysed gives the designer some idea of the final reliability of his creation, and how much work is needed to attain the specified figure. Only by testing products until they fail, investigating the failures and designing them out, will reliable equipment be produced.

When development is completed, the item goes for manufacture, and it becomes important to ensure that the item made on the production line is identical to the one that resulted from the development. To this end, some degree of inspection, quality control and quality assurance is necessary, although it must be said here that reliability cannot be inspected into a product. It must be designed in.

The final phase of a product is when it is being used — and maintained, updated and no doubt abused. The most valuable activity that can go on at this time is the collection of data. This is probably the most difficult. This book does not discuss the

16   Introduction

establishment and running of reliability data bases, nor the difficulties involved in persuading tired mechanics and disillusioned users to properly report the details of the failures they experience. Sufficient to say it is a management problem of large proportions.

For example, imagine that we have to design a washing machine. Initially the price, the investment that would be needed in new plant, the type of machine — upright or front loader, automatic or manual — would be considered. During the feasibility study broad aspects of different designs would be examined, for example the belt or gear driven from the electric motor, should the drum counter-rotate, electronic control for an automatic, how is the water heated (gas, electricity or is the machine to be connected to the household hot water supply), and last, but by no means least, how reliable is it to be.

Suppose we have decided on a front-loader, automatic, with an electronic control. The electric motor is to have a belt drive to the drum, and will counter rotate during the wash cycle (to prevent the clothes becoming tangled) by reversing the polarity of the current through the motor. A variable current is to be used to control the speed. The machine will have a hot and cold water input, and an immersion heater and a thermostat will be incorporated (a fairly luxurious machine). There will be a maintenance policy, to cover the terms of the guarantee, and also to be sold when the guarantee has expired. The reliability of any machine that has been properly maintained will be 95% over one year (i.e. no more than 5% of the machines sold are to require any maintenance, other than preventative maintenance, during each year of life). The scheduled maintenance will take place once a year, and should take no more than half an hour. Repair or unscheduled maintenance will take no more than 1 h on 50% of the times that it is needed, and no more than 2 h on 20% of them. The company should have some idea of the use the machines are likely to experience in a year. Data may exist on this point, but market research may have to be done.

Prototypes of our washing machine must be tested so that we can collect reliable data. Tests on the whole unit, as well as on the individual components, are to be recommended as failures may not be independent. For example, certain failings in the control system may well cause extra stresses to the motor and bearings, hence shortening their lives. The tests should take the possible environments into account, as well as the operator training (minimal in this case, as it is with most household equipment) and possible overstressing, e.g. what if the machine is overloaded, or left switched on, or too much detergent is used? This data is used to find system weaknesses, which can then be designed out. This test, analyse and fix regime leads to reliability growth, and should be applied to every project. It is important that the equipment being developed is thoroughly tested, as the object is to find design weaknesses to be removed. Indeed, there is a school of thought that believes the equipment *should* be overstressed during this process, to stimulate failures that are indicative of the design weaknesses. Certainly reliability trials should be distinct from the other performance trials, as during the latter failures are to be avoided if at all possible.

It may be that a reliability demonstration test is required at the end of development and before production starts, but there are disadvantages to such a test. First, it can be extremely costly, particularly as it needs expensive equipment which must be highly reliable. For example, the author was once asked to advise on the conduct of a proposed trial to demonstrate a reliability of 99.8 % for a pyrotechnic device. Something like 2000 of the devices would have to be tested (destructively) in order for the results to be statistically valid. As they cost £500 each, it would have been a very expensive trial!

Second, everybody, including the customer, wants the trial to be a success. What is to be done if the criteria for acceptance are not met? Indeed, deciding on such criteria can be a major effort in itself. The embarrassment and stress caused by a possible rejection very often leads to the equipment being non-representative so as to assure acceptance, and the trial becomes a very expensive showpiece. What is more, the money spent does not go towards improving the reliability one iota. Reliability demonstration tests are not to be recommended.

Once the design is settled, precautions must be taken to ensure that the potential reliability of the design is met during production. Production reliability assurance testing (PRAT), quality control and quality assurance (QA) can be used to help the production engineer during manufacture. It is beyond the scope of this book to describe these techniques, apart from a brief introduction to QA. But it is essential to ensure that the reliability that is designed in is also manufactured in.

2.3 References

There are many books and papers written on reliability and related topics, from a number of points of view. A few of the better known are listed here, and referenced in later chapters, though it must be said that this list is by no means exhaustive; rather the contrary, as to keep the number of references reasonable, the list has been kept short. Some of the standards are also mentioned here, as anybody who wishes to become involved in reliability must become acquainted

18   Introduction

with these. Needless to say, some of these standards are for the defence industry.

1. BAIN, L. J. *Statistical analysis of reliability and life testing models, theory and methods*, Marcel Dekker Inc. (1978)
2. BARLOW, R. E. and PROCHAN, F. *Statistical theory of reliability and life testing*, Holt, Reinhart and Winston, Inc. (1981)
3. BERGER, J. O. *Statistical decision theory and Bayesian analysis* (Second edition), Springer-Verlag (1985)
4. CARTER, A. D. S. *Mechanical reliability* (Second edition), Macmillan (1986)
5. DEGROOT, M. H. *Probability and statistics*, McGraw-Hill (1975)
6. DEGROOT, M. H. *Optimal statistical decisions*, McGraw-Hill (1970)
7. GNEDENKO, B. V., BELYAYEU, Y. K. and SOLOVYEU, A. D. *Mathematical methods of reliability theory*, Academic Press (1969)
8. KALBFLEISCH, J. D. and PRENTICE, R. L. *The statistical analysis of failure time data*, Wiley (1980)
9. LLOYD, D. K. and LIPOW, M. *Reliability: management, methods and mathematics*, Prentice Hall (1962)
10. MANN, N. R., SCHAFER, R. E. and SINGPURWALLAH, N. D. *Methods for statistical analysis of reliability data*, Wiley (1974)
11. MARTZ, H. F. and WALLER, R. A. *Bayesian reliability analysis*, Wiley (1982)
12. O'CONNOR, P. D. T. *Practical reliability engineering*, Wiley (1985)
13. OTT, L. *An introduction to statistical methods and data analysis*, Duxbury Press (1977)
14. *British Standard BS 5750 Quality systems*
15. *British Standard BS 5760 Reliability of systems, equipments and components*
16. *British Standard BS 6001 Sampling procedures and tables for inspection by attributes*
17. *Def. Stan. 00-40 The management of reliability*
18. *Def. Stan. 00-41 MOD practices and procedures for reliability and maintainability*
19. *US MIL-HDBK-189 Reliability growth management*
20. *US MIL-HDBK-217E Reliability prediction for electronic systems*
21. *US MIL-STD 1629 Failure modes and effects analysis*
22. *US MIL-STD 1635 Reliability growth testing*

Chapter 3

Reliability as a function of time

Because reliability is a probability, as defined in the previous chapter, and because time is a part of that definition, it is clear that the life of the equipment will vary in a random manner from one equipment to the next. The problem of dealing with this randomness must then involve the use of statistics. Indeed, statistical techniques are necessary to analyse reliability data at all stages in the development and life of an equipment. To predict reliability from information on subassemblies, to analyse data on times to failure, making decisions on maintenance policies, all need a knowledge of statistics. Thus although reliability is not a statistical problem, statistics forms an integral part of any serious work on the estimation or prediction of reliability.

3.1 Probability

Consider the following situations:

1. A reliability engineer examines 100 bearings, and finds 11 of them are out of tolerance. His conclusion is that the probability of a bearing being out of tolerance is about 0.11 or 11%, meaning that if a large number of similar bearings were to be examined, about 11% of them would be found to be out of tolerance. Thus probability is defined to be long term relative frequency.
2. If a balanced coin is tossed, the probability that it lands head uppermost is 0.5. We expect that about 50% of a large number of throws will be heads.
3. The reliability of a washing machine in its first year of use is 95%, or the probability of the machine not breaking down in its first year is 0.95. The manufacturer knows that 95% of the machines he produces will not break down during their first year.

Probability is usually estimated from data, rather than from a knowledge of the symmetries of a situation (as in the case of the coin). Some of the elementary ideas of statistical analysis are dealt with in Chapter 6. More advanced ideas, such as sampling theory, design of

20 Reliability as a function of time

experiments and other statistical techniques, will not be dealt with in this text. However, there is a need to revise the laws of probability. When estimating the reliability of a complex system, such as an aircraft or an oil platform, it is impractical to test a large number of the completed items to destruction, and the reliability of the system has to be estimated from the reliability of its components and subsystems. There are several rules which apply:

1. *The product rule.* If A and B are two events of interest, and P_a and P_b the respective probabilities of them occurring, then if A and B are independent (i.e. knowledge about whether or not one of them has occurred does not affect the probability of the other happening), then

$$P\,(A \text{ and } B \text{ happen}) = P_a \times P_b$$

2. *The summation rule.* If A, B, P_a, P_b are defined as above, and A and B are mutually exclusive (i.e. they cannot both happen), then

$$P\,(A \text{ or } B \text{ happens}) = P_a + P_b$$

3. If $A_1, A_2, \ldots, A_n$ are mutually exclusive events, and they describe all possible outcomes in a particular situation, then

$$P_1 + P_2 + \cdots + P_n = 1$$

4. In particular, if there are only two possible outcomes in a particular situation, call them success and failure, then

$$P\,(\text{success}) = 1 - P(\text{failure})$$

3.2 Component and non-repairable systems reliability

Consider the data presented below in Table 3.1. This is the time to failure of 100 generators that were run until failure on a test.
f_i is the number of generators that failed during month i, F_i is the total number that had failed by the end of month i, and R_i is the total number that were still functioning at the end of month i. So

$$F_i = \sum_{j=1}^{i} f_j$$

$$R_i = 100 - F_i$$

The data can be presented as a histogram as in Figure 3.1 or as a cumulative frequency as in Figure 3.2.

If R_i is divided by 100, the total number of items on test, the result is the proportion of items still functioning at the end of the ith month, or the reliability at the end of the ith month. So the reliability is 92 %

Table 3.1

Month	f_i	F_i	R_i	λ_i
1	1	1	99	0.010
2	0	1	99	0.0
3	2	3	97	0.020
4	1	4	96	0.010
5	4	8	92	0.040
6	10	18	82	0.109
7	18	36	64	0.220
8	22	58	42	0.344
9	14	72	28	0.333
10	8	80	20	0.286
11	10	90	10	0.500
12	5	95	5	0.500
13	3	98	2	0.600
14	1	99	1	0.500
15	0	99	1	0.0
16	1	100	0	1.0

at the end of the fifth month, for example. Similarly, F_i divided by 100 gives the probability of failure by the end of the ith month. For example, the probability that a pump has failed by the end of the third month is 18%.

The last column, λ_i, is related to the others by the equation

$$\lambda_i = f_i/R_{i-1}$$

i.e. it is the number of failures in a given month as a proportion of the number still functioning at the beginning of the month, and so is the probability of an item failing in month i, assuming it is still

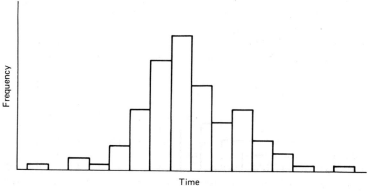

Figure 3.1 Histogram of times to failure

22 Reliability as a function of time

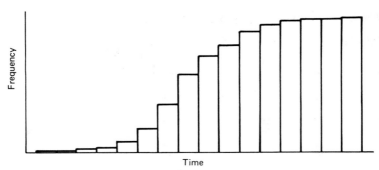

Figure 3.2 Cumulative frequency of times to failure

functioning at the beginning of the month. Figure 3.3 shows the graph of λ_i for the data above. λ_i is called the failure rate or hazard rate.

It is not unusual to assume that there are underlying mathematical formulae describing the functions f_i, F_i, R_i and λ_i. In that case their definitions and the relationships between them are set out below.

1. *The probability density function*, f(t). This is the way a statistician might first think of using to describe the data. It is, however, not necessarily the best way of presenting the data for our purposes. It has the following property. If t is the time to failure, and a and b are two times, then:

$$P(t \text{ lies between } a \text{ and } b) = P(a < t < b)$$
$$= \int_a^b f(t)\,dt$$

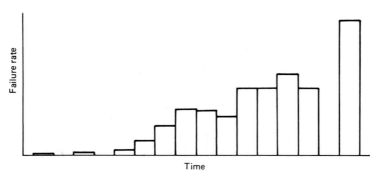

Figure 3.3 Failure rate

2. *The reliability*, $R(t)$. This is related to $f(t)$ by the equation

$$R(t) = \int_t^\infty f(t)\,dt$$

or

$$f(t) = -dR/dt$$

$R(t)$ is the probability that the item is still functioning by time t, or alternatively that it fails after time t.

3. *The failure function*, $F(t)$. The relationships are

$$F(t) = \int_0^t f(t)\,dt$$
$$= 1 - R(t)$$

and

$$f(t) = dF/dt$$

4. *The failure rate.* This is given by

$$\lambda(t) = f(t)/R(t)$$

and has the property that if an item has not failed at time t, then the probability that it fails in the interval t to $t + dt$ is $\lambda(t)\,dt$. If $\lambda(t)$ is known, R can be derived using the formula

$$R(t) = \exp - \int_0^t \lambda(t)\,dt$$

The one parameter (as opposed to function) often used when describing reliability is the mean time to failure or MTTF. This is just the average lifetime of a large number of the items, and is

$$MTTF = \int_0^\infty tf(t)\,dt$$
$$= \int_0^\infty R(t)\,dt$$

3.3 The bathtub curve

Figure 3.4 below shows a typical failure rate curve (the failure rate as a function of time). Because of its shape it is usually referred to as the 'bathtub' curve.

Initially the failure rate is high, as undetected manufacturing faults make themselves felt. In practice, burn-in is often used to remove

24 Reliability as a function of time

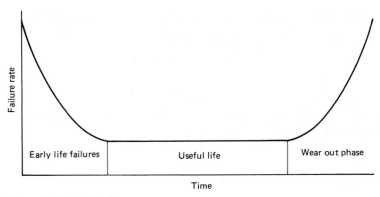

Figure 3.4 The bathtub curve

these faulty items. The failure rate is decreasing during this period, and is known as decreasing failure rate (DFR). This is followed by a period of constant failure rate (CFR), known as the useful life. Finally the item reaches the wear out period, and has increasing failure rate (IFR), and should be discarded. In principle we are describing a non-repairable item in this section, but complex, repairable items also exhibit these features. Two examples are discussed below.

The mortality of human beings fits the bathtub curve very well. Indeed it was actuaries, interested in problems of life insurance, who first studied this curve. Initially humans suffer a relatively high mortality rate, as those who are born with some defect perish. For example, it is very difficult to insure the life of a young child. (A medical man was once heard to say at a conference 'The first year of life is the most dangerous.' The reply was that the last is not without its problems.) Then humans enter a period of relative safety, when death is from accident and non-age-related diseases, until by late middle age they enter the time when they are at risk from such things as diabetes and heart disease that increase the failure rate.

Cars also exhibit the bathtub behaviour. A car in the first year of its life requires more attention than one a little older, as faulty items are replaced. Then in the useful life period, it is less likely to fail, certainly if it is properly maintained, until in old age it begins to fall apart faster than the owner can afford to have it repaired.

Program 3.1 calculates R and λ as a function of time. The data input by the user are the number of failures in each time period (hour, day, month etc.). It assumes that all the items were new at the start of the first time period.

Models for failure rate   25

```
100 REM This program calculates estimates of the Reliability
110 REM and failure rate from inputted data. The inputs are
120 REM the number of failures in each time period (each
130 REM hour or month etc.)
140 PRINT " Input the number of readings";
150 INPUT M
160 DIM F(M),N(M),P(M)
170 PRINT " Now enter the number of failures in each period"
180 FOR I= 1 TO M
190 PRINT "Enter the number of failures in period";I;
200 INPUT F(I)
210 TOTAL=TOTAL+F(I)
220 NEXT I
230 PRINT
240 FOR I=1 TO M
250 N(I)=TOTAL-F(I)
260 P(I)=F(I)/TOTAL
270 TOTAL=TOTAL-F(I)
280 NEXT I
290 REM
300 PRINT " Period      Number of      Number      Proportion"
310 PRINT "             failures      Surviving      failing"
320 PRINT
330 FOR I=1 TO M
340 PRINT I,F(I),N(I),P(I)
350 NEXT I
360 END
```

Program 3.1 Calculating estimates of the reliability and failure rates

3.4   Models for failure rate

A mathematical expression is very often used to describe the reliability functions. Some of these are described below.

Constant failure rate

If we assume that λ is constant with time, then we get

$$R(t) = \exp(-\lambda t)$$
$$f(t) = \lambda \exp(-\lambda t)$$
$$F(t) = 1 - \exp(-\lambda t)$$

and

$$\mathrm{MTTF} = 1/\lambda$$

This particular model has a number of advantages and disadvantages. The advantages include:

1. ease of algebraic manipulation;
2. easy estimation of λ $(= 1/\mathrm{MTTF})$;

26   Reliability as a function of time

3. a reasonable description of certain systems, in particular complex repairable systems in their useful life period and under certain circumstances certain electronic components.

The disadvantages are:

1. with only one parameter to vary, the data is not always a good fit to the model;
2. it assumes the item does not age, i.e. the probability of failure in the interval from time t until time $t + x$ depends only on x, the length of the interval, and not on t, the age of the item.

This last statement certainly does not apply to many components, particularly mechanical items that are subject to wear. However, there are more complex models discussed below.

The Weibull distribution

For this model we put λ proportional to a power of t.

$$\lambda(t) = \frac{\beta}{\gamma} \left(\frac{t}{\gamma} \right)^{\beta - 1}$$

$$R(t) = \exp - \left(\frac{t}{\gamma} \right)^{\beta - 1}$$

$$f(t) = \frac{\beta}{\gamma} \left(\frac{t}{\gamma} \right)^{\beta - 1} \exp - \left(\frac{t}{\gamma} \right)^{\beta - 1}$$

where:
    β = the shape parameter
    γ = the scale parameter, or characteristic life.
Note that when $t = \gamma$, then $R = 0.32$.

There are good reasons for using this distribution, some analytic, some heuristic:

1. the simplest function that is more complex than λ constant is λ proportional to a power of t
2. a two-parameter model is more flexible than the simple CFR case described above, and far more data sets give a reasonable fit to the Weibull distribution
3. statistical theory tells us that the Weibull distribution describes the 'weakest link' very well. As many items fail because they have some weaknesses due to the manufacturing process, they fail when the 'worst' of these manifests itself.

Graphs of λ, f and R are shown in Figure 3.5 for different values of

Models for failure rate 27

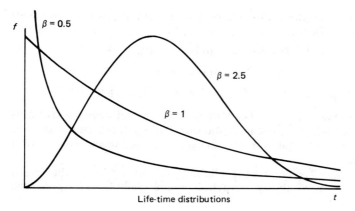

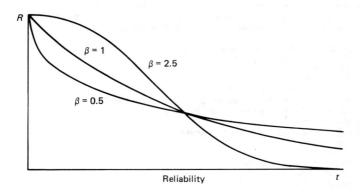

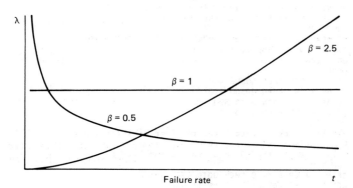

Figure 3.5 The Weibull curve

28   Reliability as a function of time

β. Notice that the behaviour of λ varies qualitatively for different values of β. In fact

$\beta < 1$ The failure rate is decreasing (DFR)

$\beta = 1$ The failure rate is constant (CFR)

$\beta > 1$ The failure rate is increasing (IFR)

Also observe that for large values of β, the reliability R remains relatively large (i.e. close to 1) and then falls off dramatically as t gets close to γ. Techniques for estimating β and γ are discussed in a later chapter.

Program 3.2 calculates the reliability at a given time, or the time at which a given reliability is reached for the Weibull and exponential distributions.

```
100 REM This program calculates the Reliability at a given
110 REM Time or the Time at which a given Reliability is
120 REM achieved, for the exponential or Weibull distribution.
130 PRINT " Input E(exponential) or W(Weibull)"
140 INPUT D$
150 IF D$="E" OR D$="W" THEN GOTO 180
160 PRINT " Wrong input. Try again."
170 GOTO 130
180 SHAPE = 1
190 IF D$="E" THEN GOTO 260
200 PRINT " Input the parameters of the distribution"
210 PRINT " The scale parameter first";
220 INPUT MTBF
230 PRINT " and now the shape parameter";
240 INPUT SHAPE
250 GOTO 280
260 PRINT " Input the failure rate";
270 INPUT MTBF
280 PRINT
290 PRINT " Enter T for the time at which the reliability is R"
300 PRINT " Enter R for the reliability at time t"
310 PRINT " Enter A for a change of distribution"
320 PRINT " Enter Q to quit the program"
330 INPUT A$
340 IF A$="T" OR A$="R" OR A$="Q" OR A$="A" THEN GOTO 370
350 PRINT " Wrong input. Try again."
360 GOTO 280
370 PRINT
380 IF A$ = "T" THEN GOSUB 430
390 IF A$ = "R" THEN GOSUB 510
400 IF A$ = "Q" THEN GOTO 590
410 IF A$ = "A" THEN GOTO 130
420 GOTO 280
430 REM This section calculates the time at which a given
440 REM  reliability is achieved
450 PRINT " Enter the value of the reliability";
460 INPUT R
470 IF R > 1 OR R <= 0 THEN GOTO 430
480 LET T = (-(LOG(R)*MTBF))**(1/SHAPE)
490 PRINT "The time is";T;" ."
```

Worked examples 29

```
500 RETURN
510 REM This section calculates the reliability
520 REM at a given time
530 PRINT "Enter the time ";
540 INPUT T
550 IF T < 0 THEN GOTO 510
560 LET R = EXP(-(T/MTBF)**SHAPE)
570 PRINT "The reliability is ";R;" ."
580 RETURN
590 END
```

Program 3.2 Reliability at a given time or time to reach a given reliability

The log normal distribution

The log normal distribution is another model for failure. Here:

$$f(t) = \frac{1}{t\sigma\sqrt{2\pi}} \exp -1/2 \left(\frac{\ln(t) - \mu}{\sigma} \right)^2$$

There is no analytic formula for R or λ. Graphs of these functions are shown in Figure 3.6. The log normal distribution is often used to describe repair times.

The truncated normal distribution

In this case:

$$f(t) = K \exp -1/2 \left(\frac{t - \mu}{\sigma} \right)^2 \qquad t > 0$$

Again there is no analytic form for λ or R, and μ and σ are difficult to estimate from data, so this model is rarely used. Graphs showing λ, f and R are in Figure 3.7.

The topics discussed in this chapter are so basic to reliability that it would be impossible to pull just one of the references out of the list and recommend it. Any one of the references listed at the end of Chapter 2 with 'reliability' in the title will give some account of the topics in this chapter.

WORKED EXAMPLES

(3.1) Consider the simple telephone system shown in Figure 3.8. For a user at A to be able to speak to a user at B, both phones and the exchange must function properly. If the reliability of each of the phones is 97 %, and that of the exchange is 95 %, then the reliability of

30 Reliability as a function of time

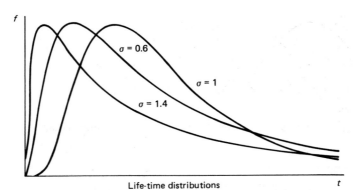

Life-time distributions

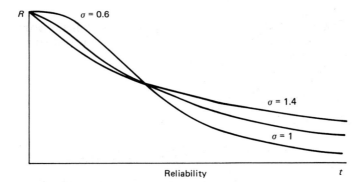

Reliability

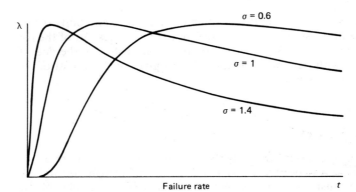

Failure rate

Figure 3.6 The lognormal distribution

Worked examples 31

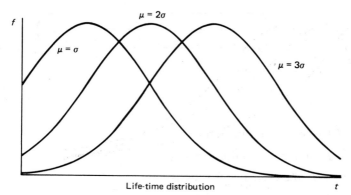

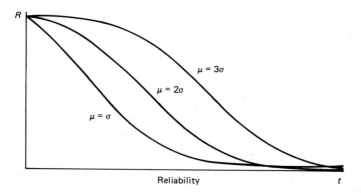

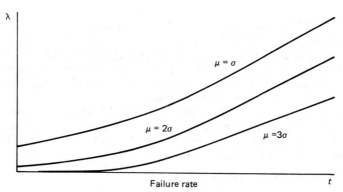

Figure 3.7 The truncated normal distribution

32 Reliability as a function of time

Figure 3.8 Simple telephone system

the system
$$\begin{aligned} &= P \text{ (the system functions)} \\ &= P \text{ (phone A functions)} \times P \text{ (phone B functions)} \\ &\quad \times P \text{ (the exchange functions)} \\ &= 0.97 \times 0.97 \times 0.95 \\ &= 0.894 \end{aligned}$$

(3.2) In a quality assurance scheme a sample of 100 items from a large batch is examined. The batch is considered acceptable if two or fewer items are found to be defective. For a given proportion, p, of defectives in the batch let p_i be the probability that exactly i defectives are found in the sample. Then the probability that a batch is not sentenced as defective

$$\begin{aligned} &= P \text{ (2 or fewer defectives are found)} \\ &= p_0 + p_1 + p_2 \end{aligned}$$

(Quality assurance will be dealt with in a later chapter.) In this example i can take values $0, 1, 2, \ldots, 100$ and none other. Recall that:

$$p_0 + p_1 + p_2 + \cdots + p_{100} = 1$$

(3.3) The equation P (success) $= 1 - P$ (failure) is very often used in reliability calculations, when the unreliability of a device is calculated although its reliability is wanted. Consider the telephone system shown in Figure 3.9, which is similar to that discussed above, but with the addition of a standby exchange.

The system functions if both the phones function and the exchange subsystem functions. The exchange subsystem functions if one or both of the exchanges function, or turning the problem round, the subsystem fails if both of the exchanges fail.

$$\begin{aligned} &P \text{ (the exchange system functions)} \\ &= 1 - P \text{ (the exchange system fails)} \end{aligned}$$

Worked examples 33

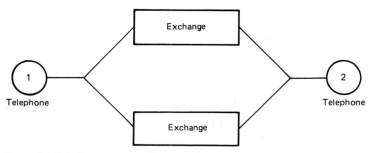

Figure 3.9 Telephone system with redundancy

and

P (the exchange system fails)

$= P$ (both the exchanges fail)

$= P$ (exchange 1 fails) $\times P$ (exchange 2 fails)

$= (1 - 0.95) \times (1 - 0.95)$

$= 0.05 \times 0.05$

$= 0.0025$

P (the exchange system functions)

$= 0.9975$

Proceeding now as in Example **(3.1)**

P (the system functions)

$= 0.97 \times 0.97 \times 0.9975$

$= 0.939$.

(3.4) Five electronic components were subjected to stress until they failed. The observed failure times were (in 100 h)

5.5, 6.8, 8.2, 9.1, 11.1

Calculate the MTTF, the failure rate, the reliability at 1000 h and the time at which 15% of a large number of similar components subjected to similar stresses could be expected to have failed.

34 Reliability as a function of time

Solution

$$\text{MTTF} = (5.5 + 6.8 + 3.2 + 9.1 + 11.1)/5$$
$$= 8.14$$
$$\lambda = 1/\text{MTTF}$$
$$= 0.123 \text{ failures per } 100 \text{ h}$$
$$R(1000) = \exp - 10/8.14$$
$$= 0.29$$

To find the time at which 15% will have failed, remember that

$$F(t) = 1 - R(t)$$

and so

$$0.15 = 1 - \exp - t/8.14$$
$$0.85 = \exp - t/8.14$$

Taking natural logs,

$$t/8.14 = -\ln(0.85)$$
$$t = 8.14[-\ln(0.85)]$$
$$= 1.32 \times 100 \text{ h}$$
$$= 132 \text{ h}$$

(3.5) Twenty components with CFR were observed being used in a high stress environment. After 25 h use, seven of them had failed at times (in hours)

$$2.1, 8.3, 10.9, 15.2, 16.3, 20.5, 23.8$$

while the remaining 13 were still functioning. Calculate the MTTF, the failure rate, the time at which the reliability is 95% and the reliability after 50 h of use.

Solution

The MTTF in a case like this is the sum of all the test times divided by the number of failures. This gives

$$\text{MTTF} = [2.1 + 8.3 + 10.9 + 15.2 + 16.3 + 20.5$$
$$+ (13 \times 25)]/7$$

The first seven terms of the above expression are the failure times of the failed items, but the total time on test must include the thirteen unfailed items which each contribute a test time of 25 h, which

Worked examples 35

accounts for the last term of (13×25). This data is known as censored (as we don't know the failure times of 13 of the components).

$$\text{MTTF} = 422.1/7$$
$$= 60.3$$
$$\lambda = 1/60.3$$
$$= 0.0166 \text{ failures/h}$$
$$0.95 = \exp - t/60.3$$
$$t = 60.3 - \ln(0.95)$$

as before

$$= 3.1 \text{ h}$$
$$R(50) = \exp - 50/60.3$$
$$= 0.44$$

(3.6) An item is known to have a failure time that is Weibull distributed with characteristic life 250 h and shape parameter 2.5. What is the reliability at 100 h and at what time is the reliability 95%?

$$R(100) = \exp - \left(\frac{100}{250}\right)^{2.5}$$
$$= 0.904$$
$$R(t) = 0.95$$
$$= \exp - \left(\frac{t}{250}\right)^{2.5}$$

Taking logs of both sides,

$$\left(\frac{t}{250}\right)^{2.5} = \ln(0.95)$$
$$= 0.051$$
$$t = 250 \times (0.051)^{1/2.5}$$
$$= 76.2 \text{ h}$$

Note that in this case, $\text{MTTF} = 222$ h. (It is beyond the scope of this book to go further into why this is so.)

36 Reliability as a function of time

(3.7) An item with CFR and MTTF of 222 h has the following failure characteristics:

$$R(100) = \exp - 100/222$$
$$= 0.637$$

and when

$$R(t) = 0.95$$
$$= \exp - t/222$$

$t = -222 \times \ln(0.95)$

$$= 11.4 \, \text{h}$$

In other words, the reliability falls off much more rapidly in the CFR case. The difference in the two results demonstrates the importance of modelling with the correct distribution.

PROBLEMS

(3.1) Two hundred and fifty bearings were observed in use for two years, and the data recorded, as shown in Table 3.2.

Table 3.2

Month	Number of failures	Month	Number of failures
1	13	13	2
2	9	14	2
3	7	15	1
4	6	16	2
5	4	17	3
6	3	18	2
7	3	19	1
8	2	20	2
9	2	21	3
10	2	22	4
11	1	23	6
12	2	24	10

Use Program 3.1 to calculate the reliability, the failure probability and the failure rate for each month from 1 to 24.

(3.2) Eight components on a cross-country vehicle were tested to destruction, and the distances to failure were (in 1000 km)

$$5.6, 6.9, 8.2, 10.3, 13.2, 16.7, 19.7, 24.6$$

Worked examples   37

Find the mean distance to failure (MDTF), and failure rate, the reliability at 500 km and the distance at which the reliability is 80%.
(*Solution:* 13.15 km, 0.076 failures/100 km, 96%, 2934 km.)

(**3.3**) Fifteen microswitches were put on test, and the number of cycles until failure counted. (A cycle constitutes switching a switch on and then off again.) After 10 000 cycles six of them had failed. The number of cycles at failure were

$$2924, 3722, 4209, 6039, 7179, 8357$$

Calculate the mean number of cycles to failure, the failure rate, the number of cycles at which the reliability is 95% and the reliability at 4000 cycles.
(*Solution:* 8162, 0.00012, 419, 61%.)

(**3.4**) A component is known to have a failure time that is Weibull distributed with shape parameter 0.5 and scale parameter 150 h. What is the reliability at 50 h and at what time is the reliability 90%?
(*Solution:* 0.56, 1.67 h.)

The MTTF is 300 h in this case. Compare your result with the case of a component with CFR and MTTF of 300 h.
(*Solution:* 0.84, 31.6 h.)

Chapter 4
Systems modelling

The previous section examined the reliability of individual components. This section examines the consequences of putting together a number of components into a complex piece of equipment which we call a system. It is often very difficult and/or expensive, if not actually impossible, to test a sufficient number of systems to obtain results that have sufficient statistical validity, and in these cases it is necessary to be able to calculate the systems reliability from a knowledge of the components. It is also necessary to deal with other system parameters, such as availability, maintainability and safety. This chapter discusses some theoretical techniques that can be used.

4.1 Reliability block diagrams

Consider a motor-bicycle, and in particular the subsystem consisting of the tyres. The subsystem has failed if either of its components fails. We call a system like this a series system, and represent it by the diagram in Figure 4.1.

The alternative type of system is illustrated by a two-engined aircraft that can fly on one engine. In this case there is a certain amount of redundancy, and this system is represented by the diagram in Figure 4.2.

Diagrams of this nature are used to represent complex systems as an aid when making reliability predictions at the feasibility and design stages of a program, and are called reliability block diagrams, or RBDs. They represent the reliability aspects of a system, and need not reflect its physical structure. They are best thought of in the following way. Each component or subsystem is represented by a

Figure 4.1 RBD of motorbicycle tyres

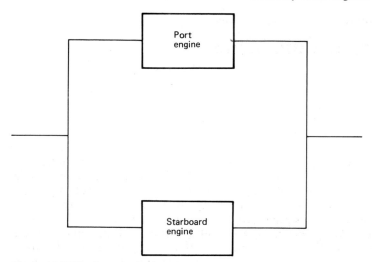

Figure 4.2 RBD of two-engined aircraft

switch that is closed if the system is functioning and open if it is failed. The switches are configured in such a way that the system functioning/failing corresponds to current flowing/not flowing from one side of the diagram to the other for a given pattern of component failures/open switches. We shall only study systems that can be broken down into the simple configurations shown above.

Series systems

These are those systems represented by the diagram shown in Figure 4.1, or if there are several components, the diagram in Figure 4.3.

In this case the system works only if all the components work, and then, if the failures are independent

$$R = R_1 \times R_2 \times R_3 \ldots R_n$$

i.e. the component reliabilities are multiplied together. In general R is

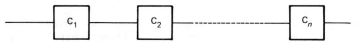

Figure 4.3 General series system

40 Systems modelling

a function of time, in which case

$$f = -dR/dt$$

$$= -\frac{dR_1}{dt} R_2 \dots R_n - R_1 \frac{dR_2}{dt} R_3 \dots R_n - R_1 R_2 \frac{dR_3}{dt} \dots$$

$$\dots R_n - R_1 R_2 R_3 \dots \frac{dR_n}{dt}$$

and

$$\lambda = f/R$$

$$= -\frac{dR_1/dt}{R_1} - \frac{dR_2/dt}{R_2} - \frac{dR_3/dt}{R_3} - \dots - \frac{dR_n/dt}{R_n}$$

$$= \frac{f_1}{R_1} + \frac{f_2}{R_2} + \frac{f_3}{R_3} + \dots + \frac{f_n}{R_n}$$

$$= \lambda_1 + \lambda_2 + \dots + \lambda_n$$

i.e. the failure rate of the system is the sum of the failure rates of its components.

Redundant systems

These are also called parallel or standby systems. There are a number of cases:

1. *Hot redundancy.* This is the situation when all the redundant items are running, although the system will function if some of them have failed. The two engined aircraft above is an example of a hot redundant system.
2. *Cold, or standby redundancy.* This is the system in which the redundant items are kept switched off until they are needed. The spare tyre on a car is a good example of standby redundancy. The advantage of standby redundancy is that items do not wear or age as much in the dormant state as those in use, while the disadvantage is that a switching system is needed to bring them up if needed. It is no use a car carrying a spare if the driver is unable to change the wheel.
3. k-*out-of*-n *systems.* In a way these are systems that are part way between series and full redundant systems. There are n components available, of which k are needed for the system to run

properly. They can be hot or cold redundant. The car's spare is an example of a four-out-of-five system, as any four of the five tyes is sufficient for the car to function.

In this book we shall only analyse 1-out-of-n hot redundant systems. The symbolism used in an RBD is shown in Figure 4.4.

The system only fails if all the components fail. Then

$$F = F_1 \times F_2 \ldots F_n$$

or, after doing a little algebra,

$$R = 1 - (1 - R_1) \times (1 - R_2) \ldots (1 - R_n)$$

In general there is no easy relationship between the failure rate λ of the system and the failure rates λ_i of the components. For example, consider the case when there are just two components in parallel, each with CFR. Then

$$R_i = \exp(-\lambda_i t) \qquad i = 1, 2$$

and

$$R(t) = 1 - (1 - R_1)(1 - R_2)$$
$$= \exp - \lambda_1 t + \exp - \lambda_2 t - \exp - (\lambda_1 + \lambda_2)t$$

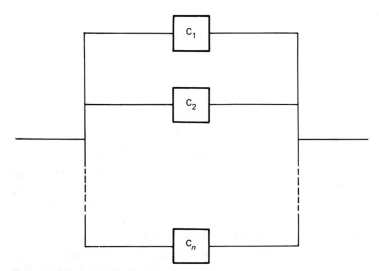

Figure 4.4 General redundant system

$$f(t) = -\frac{dR}{dt}$$
$$= \lambda_1 \exp - \lambda_1 t + \lambda_2 \exp - \lambda_2 t - (\lambda_1 + \lambda_2)\exp - (\lambda_1 + \lambda_2)t$$
$$\lambda(t) = f/R$$
$$= \frac{\lambda_1 \exp - \lambda_1 t + \lambda_2 \exp - \lambda_2 t - (\lambda_1 + \lambda_2)\exp - (\lambda_1 + \lambda_2)t}{\exp - \lambda_1 t + \exp - \lambda_2 t - \exp - (\lambda_2 + \lambda_2)t}$$

Notice that this is not constant with time. However,

$$\lambda(t) \to \min(\lambda_1, \lambda_2) \quad \text{as} \quad t \to \infty$$

A graph of $\lambda(t)$ is shown in Figure 4.5.

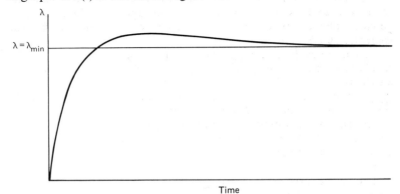

Figure 4.5 Failure rate function of a hot redundant system

$$\text{MTTF} = \int_0^\infty tf(t)\,dt$$
$$= 1/\lambda_1 + 1/\lambda_2 - 1/(\lambda_1 + \lambda_2)$$
$$= 2/3\lambda \text{ in the special case when } \lambda_1 = \lambda_2 = \lambda \text{ (i.e. when the components are identical)}$$

MTTF = 3/2 (MTTF) of each of the components)

Thus in the case when the components are identical, the MTTF is increased by 50% by putting in a (hot) redundancy.

Further worked examples are shown at the end of the chapter.

4.2 Maintainability, availability and replacement

Having discussed how to describe the reliability of a component in the previous chapter, we must now discuss the characteristics of a

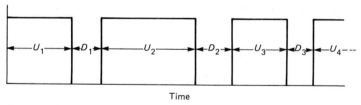

Figure 4.6 Up times and down times of complex equipment

repairable item, such as our washing machine. Consider the graph illustrating the up times and down times of an equipment shown in Figure 4.6. (The up time is when the equipment is functioning, while the down time is when it has failed and is being repaired or is waiting repair.) The up times are $U_1, U_2, \ldots$ and the down times are $D_1, D_2, \ldots$.

In general each of the up and down times will be random. Let $f(t)$ be the probability density function (p.d.f.) of the up times, and let $m(t)$ be the p.d.f. of the down times. From an engineering point of view, this assumes that after each repair the equipment is 'as good as new' and that there is no preventative maintenance. This is true, or very nearly so, by the time we have reached the 'useful life' part of the bathtub curve. If there were preventative maintenance, we would know, or at least have a very good idea, of some of the up and down times.

The characteristics of the up times were described in Chapter 2, under the heading of component and system reliability. It is important to know the ratio of up time to down time, and how long a down time is likely to be.

Maintainability

For the maintenance and repair times there are formulae that correspond to those given in Chapter 2 for times to failure. Let

$$M(t) = \int_0^t m(s)\,ds$$

This is defined as the maintainability of the system, and is the probability that the system is repaired by time t. The repair rate is defined as

$$\mu(t) = m(t)/(1 - M(t))$$

and is analogous to the failure rate. If μ is known then

$$M(t) = 1 - \exp - \int_0^t \mu(t)\,dt$$

44 Systems modelling

and there is also the relationship

$$m(t) = \mathrm{d}M/\mathrm{d}t$$

The mean time to repair is

$$\mathrm{MTTR} = \int_0^\infty tm(t)\,\mathrm{d}t$$

$$= \int_0^\infty (1 - \mathrm{M}(t))\,\mathrm{d}t$$

In practice, the log normal distribution is often taken to describe repair times, i.e.

$$m(t) = \frac{1}{t\sigma\sqrt{2\pi}}\exp-1/2\left(\frac{\ln(t)-\mu}{\sigma}\right)^2$$

There is no analytic expression for M, or the repair rate. However, if a number of repair times

$$m_1, m_2, \ldots, m_n$$

then it can be shown that the best estimate of μ is given by

$$\hat{\mu} = \sum \ln(m_i)/n$$

$$= \ln(n\sqrt{\Pi m_i})$$

i.e. the natural log of the geometric (as opposed to the arithmetic) mean of the data.

σ is a measure of the variability in the repair time. For non-modular equipment, a value as high as 1.2 may be achieved, while for modular equipment, it may be as low as 0.6.

Availability

Availability is defined as the proportion of time that a system is functioning taken over a long time period. It is possible to show that for most patterns of failure and repair:

$$A = \frac{\mathrm{MTBF}}{\mathrm{MTBF} + \mathrm{MTTR}}$$

It is very often availability rather than reliability that the user (or customer) wants. This point is discussed in more detail in the section on specification.

Replacement

A question that may well be asked is 'How many spares will be needed?' The answer depends on the time interval concerned and the degree of risk that is acceptable, but in the constant failure rate case the following result is useful. If

$$R(t) = \exp - \lambda t$$

then the probability, P, of exactly n replacements being needed in the time interval T is given by the Poisson distribution with mean λT, i.e.

$$P_n = \frac{e^{-\lambda T}(\lambda T)^n}{n!}$$

The probability of needing n or fewer replacements is

$$P_0 + P_1 + \cdots + P_n$$
$$= e^{-\lambda T}\left(1 + \lambda T + \frac{(\lambda T)^2}{2!} + \cdots + \frac{(\lambda T)^n}{n!}\right)$$

So the probability of needing more than n spares (i.e. of running out) is

$$P = 1 - e^{-\lambda T}\left(1 + \lambda T + \frac{(\lambda T)^2}{2!} + \cdots + \frac{(\lambda T)^n}{n!}\right)$$

Values of this expression can be tabulated, or calculated on a computer, or Poisson probability paper can be used, as shown in Figure 4.7. This shows P as a function of λT, drawn on non-linear graph paper, where P is the probability of running out of spares, and a is λT, the mission time divided by the MTTF.

Programs 4.1 and 4.2 solve different aspects of this problem. The first calculates the probability of running out of spares, for a given MTTF, mission time and number carried, while the second calculates the number of spares required to meet a given risk.

```
10 REM This routine calculates the probability of running
20 REM out of spares in a given time period. It assumes CFR
110 REM   Enter the data
120 PRINT
130 PRINT "Enter the failure rate";
140 INPUT FR
150 IF FR > 0 THEN GOTO 180
160 PRINT " The failure rate must be positive. Try again"
170 GOTO 120
180 PRINT
190 PRINT "Enter the required uptime";
200 INPUT UPTIME
210 IF UPTIME > 0 THEN 240
```

46   Systems modelling

```
220 PRINT " The uptime must be positive. Try again"
230 GOTO 190
240 PRINT
250 PRINT "Enter the number of spares ";
260 INPUT RENEWAL
270 IF RENEWAL > 0 THEN 300
280 PRINT " It must be positive. Try again"
290 GOTO 250
300 REM.The calculation
310 LET TERM=1
320 LET SUBTOT=1
330 FOR I = 1 TO RENEWAL
340 LET TERM = (TERM*FR*UPTIME)/I
350 LET SUBTOT=SUBTOT+TERM
360 NEXT I
370 LET ANSWER = 1-( (EXP(-FR*UPTIME))*SUBTOT )
380 REM *** OUTPUT ANSWER ***
390 PRINT
400 IF ANSWER <= 0 THEN GOTO 440
410 IF ANSWER >= 1 THEN GOTO 460
420 PRINT "The probability of requiring more than";RENEWAL;
425 PRINT " spares is";ANSWER;" ."
430 GOTO 450
440 PRINT " *** The probability is so small it is almost zero *** "
450 GOTO 470
460 PRINT " *** The probability is so great it is almost one *** "
470 PRINT
480 END
```

Program 4.1 Probability of running out of spares

```
100 REM This routine calculates the number of spares needed
110 REM to meet a given risk of running out. The program
120 REM assumes that the item has a CFR.
130 REM
140 PRINT
150 PRINT " Enter the failure rate";
160 INPUT FR
170 IF FR.> 0 THEN GOTO 200
180 PRINT " The failure rate must be positive. Try again"
190 GOTO 140
200 PRINT
210 PRINT "Enter the expected uptime";
220 INPUT UPTIME
230 IF UPTIME > 0 THEN 260
240 PRINT " The uptime must be positive. Try again"
250 GOTO 210
260 PRINT
270 PRINT " Enter the acceptable probability of running"
280 PRINT "                  out of spares in the uptime";
290 INPUT PROB
300 IF PROB <= 1 AND PROB > 0 THEN 330
310 PRINT " The probability must be between 0 and 1."
320 GOTO 270
330 REM The calculation
340 TERM=1
350 ANSWER = 0
360 SUBTOT = 1
370 ANSWER = ANSWER + 1
380 TERM = (TERM*FR*UPTIME)/ANSWER
390 SUBTOT=SUBTOT+TERM
```

Degraded reliability 47

```
400 TRY = 1-( (EXP(-FR*UPTIME))*SUBTOT)
410 IF TRY <= PROB THEN 540
420 IF INT(ANSWER/1500) = (ANSWER/1500) THEN GOTO 460
430 IF INT(ANSWER/250) = (ANSWER/250) THEN PRINT " ^^^ STILL RUNNING ^^^ "
440 LAST=TRY
450 GOTO 370
460 REM Abort the calculation
470 PRINT " The answer is greater than";ANSWER;" ."
480 PRINT " Do you wish to continue? (Y/N) ";
490 INPUT REPLY$
500 IF REPLY$ = "Y" OR REPLY$ = "y" THEN GOTO 370
510 IF REPLY$ = "N" OR REPLY$ = "n" THEN GOTO 580
520 PRINT
530 GOTO 480
540 REM  Output the solution
550 IF ABS(TRY-PROB) > ABS(LAST-PROB) AND ANSWER > 1 THEN ANSWER = ANSWER-1
560 PRINT
570 PRINT "The number of spares required is";ANSWER;" ."
580 END
```

Program 4.2 Number of spares required

4.3 Degraded reliability

We have so far only looked at the situation when a system is either functioning perfectly or not functioning at all. In practice many systems, especially large complex systems, have a number of levels of functioning, from fully working through a number of levels of degradation including, finally, totally failed. It is this situation that we shall look at here, although very briefly.

Consider the following example. A system consists of a computer and two terminals. The system can be in one of three states: fully functioning, in the degraded state in which one of the terminals has failed or totally failed because the computer and/or both of the terminals have failed. Table 4.1 gives the probability that the system will change from any one of those three states into one of the others in a fixed time period (an hour, for example).

Table 4.1

Possible future states	Present states		
	Fully functioning	Degraded	Failed
Fully functioning	0.97	0.85	0.80
Degraded	0.02	0.13	0.05
Failed	0.01	0.02	0.15

48 Systems modelling

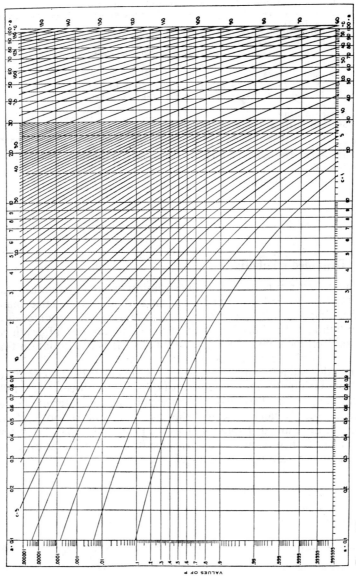

Figure 4.7

So, for example, if it is fully functioning at some time, there is a 97% chance that it will still be functioning an hour later, while if it is failed at some time, there is an 80% chance that it will be repaired and running properly an hour later. The matrix of these values

$$\mathbf{M} = \begin{pmatrix} 0.97 & 0.85 & 0.80 \\ 0.02 & 0.13 & 0.05 \\ 0.01 & 0.02 & 0.15 \end{pmatrix}$$

is called the transition matrix of the system. Notice that the columns of the matrix sum to 1, which characterizes a transition matrix. The information in the matrix can be used to calculate the proportion of time the system is in each of the three states in a long time period, and the average time it is in each of them.

Let p_1 be the proportion of time the system is in the first state, i.e. the fully functioning state. This is also the probability the system will be found in the first state after a long time. Similarly, let p_2 and p_3 respectively be the probabilities the system is in the second and third states (degraded and fully failed). The system can be represented by the diagram shown in Figure 4.8, where the three circles represent the three states, and the arrows represent the possible change of state, with the transition probabilities written next to them.

We wish to be able to calculate the values p_1, p_2 and p_3. First consider p_1. If the system is in state 1 at a particular time then it must have got to that state from one of three possible states. Either it was in state 1 previously, with probability p_1, and transition probability 0.97, or it was in state 2, with probability p_2 and transition probability 0.85, or it was in state 3, with probability p_3 and transition probability 0.80. As the probability that the system was in state 1 at the time considered is p_1, then it is a simple exercise in

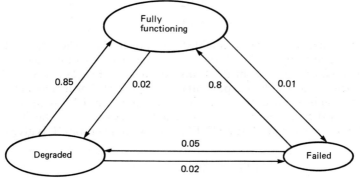

Figure 4.8 Transition diagram

50 Systems modelling

probability to obtain the result.

$$p_1 = 0.97p_1 + 0.85p_2 + 0.80p_3$$

There are similar results for p_2 and p_3, and doing a little algebra, these can be written

$$-0.03p_1 + 0.85p_2 + 0.80p_3 = 0$$
$$0.02p_1 - 0.87p_2 + 0.05p_3 = 0$$
$$0.01p_1 + 0.02p_2 - 0.85p_3 = 0$$

This relation can be written in matrix format

$$\begin{pmatrix} -0.03 & 0.85 & 0.80 & p_1 \\ 0.02 & -0.87 & 0.05 & p_2 \\ 0.01 & 0.02 & -0.85 & p_3 \end{pmatrix} = \mathbf{0}$$

or

$$(\mathbf{M} - \mathbf{I})p = \mathbf{0}$$

where $\mathbf{I}$ is the 3×3 identity matrix (with ones in the main diagonal, and zeroes elsewhere).

The p_i must also sum to 1 (as they are probabilities), i.e.

$$p_1 + p_2 + p_3 = 1$$

This gives four equations in just three unknowns. However the first three are singular, i.e. we can discard one without losing any information (the three equations sum to zero). Solving the equations in this case gives

$$p_1 = 0.965$$
$$p_2 = 0.023$$
$$p_3 = 0.012$$

i.e. the system is totally unavailable for just over 1% of the time, and is running perfectly for 96.5% of the time. There is other information that can be obtained from the transition matrix. For example, the time spent in each state is exponentially distributed. The mean of this distribution is one over the sum of the rates at which the system goes into the other states. For example, if the system is in the fully functioning state, state 1, the MTTF, or mean time in state 1, is given by

$$\text{MTTF} = 1/(0.02 + 0.01)$$
$$= 33.3 \text{ h}$$

The mean time the system is in state 2 is

$$= 1/(0.085 + 0.02)$$
$$= 1.15 \, h$$

and the mean time totally failed is

$$= 1/(0.80 + 0.05)$$
$$= 1.18 \, h$$

It is now possible to ask such questions as, what is the probability of the system being completely unavailable for longer than 1 h should it go down? Let the time be t, then

$$t = \exp - 1/1.18$$
$$= 0.43$$

using the exponential distribution.

Program 4.3 solves the equations using Gaussian elimination. The operator types in the coefficients of the transition matrix, the program subtracts one from the diagonal elements, asks the operator which of the equations is to be omitted, adds the equation expressing the fact that the probabilities sum to one, and solves these equations. The operator has the option of asking for the equations to be resolved with another equation omitted as a check, as the solution should be unchanged, entering new data, or quitting the program. The program itself has a number of checks. There are comments in the program to aid in its transcription.

```
100 REM Enter the equations
110 PRINT " Enter the number of probabilities ";
120 INPUT N
130 IF N >1 AND INT(N)=N THEN GOTO 170
140 PRINT " It must be a whole number greater than one."
150 PRINT " Try again!"
160 GOTO 110
170 Q=N+1
180 DIM M(N,Q)
190 REM This array will contain the equations
200 DIM PTOT(N)
210 REM This array checks that the coefficients for P1 to P2 add up to 1
220 FOR I=1 TO N
230 PTOT(I)=0
240 NEXT I
250 FOR I = 1 TO N
260 PRINT
270 PRINT ": Equation ";I;" :"
280 PRINT
290 FOR J = 1 TO N
300 PRINT " Enter coefficient for P";J;
310 INPUT " ";M(I,J)
320 IF M(I,J) >= 0 AND M(I,J) <= 1 THEN GOTO 360
```

52 Systems modelling

```
330 PRINT " Probabilities are between 0 and 1"
340 PRINT " Try again!"
350 GOTO 300
360 PTOT(J) = PTOT(J) + M(I,J)
370 IF PTOT(J) <= 1 THEN GOTO 410
380 PRINT " The totals are greater than one!"
390 PRINT " Input error! Try again."
400 GOTO 180
410 NEXT J
420 M(I,Q) = 0
430 REM The constant of the equations are zero
440 NEXT I
450 REM Check the totals
460 EFLAG=0
470 FOR I=1 TO N
480 IF INT(PTOT(I)+.0005)=1 THEN GOTO 500
490 EFLAG=1
500 NEXT I
510 IF EFLAG=0 THEN GOTO 550
520 PRINT " Totals of the coefficients are not one."
530 PRINT " Data error! Try again."
540 GOTO 250
550 FOR I = 1 TO N
560 M(I,I) = M(I,I) - 1
570 NEXT I
580 REM Replace an equation with one's
590 PRINT " Which equation would you like to omit (1 to ";N;")";
600 INPUT W
610 IF W >= 1 AND W <= N AND W=INT(W) THEN GOTO 640
620 PRINT " Input error! Try again."
630 GOTO 590
640 DIM EQN(Q)
650 REM Stores the equation omitted to replace it later
660 FOR I = 1 TO Q
670 EQN(I) = M(W,I)
680 M(W,I) = 1
690 NEXT I
700 REM Initialises and dimensions stores
710 DIM RESULT(N)
720 CNT = N
730 REM ...FIND FACTOR VALUES...
740 FOR F=1 TO N
750 GOSUB 1360
760 REM Position factor to be eliminated
770 GOSUB 1120
780 REM Put matrix into a temporary matrix
790 GOSUB 1210
800 REM Eliminates factors into a smaller temporary matrix
810 GOSUB 1330
820 REM Determines Xn's value
830 NEXT F
840 REM Output factors
850 PRINT
860 PRINT "* RESULTS *"
870 PRINT
880 SUM=0
890 FOR I = 1 TO N
900 PRINT " X";I;" = ";RESULT(I)
910 SUM=SUM + RESULT(I)
920 PRINT
930 NEXT I
940 IF INT(SUM+0.005) <> 1 THEN PRINT " ERROR: Total probability= ";SUM
```

The specification of reliability 53

```
950 REM  Menu - Checks, Enter more data, Quit
960 FOR I=1 TO Q
970 M(W,I)=EQN(I)
980 NEXT I
990 PRINT
1000 PRINT " ****************** MENU ****************************"
1010 PRINT " *                                                 *"
1020 PRINT " *  1. Omit a different equation to check the results *"
1030 PRINT " *  2. Enter a new set of equations                *"
1040 PRINT " *  3. Quit the program                            *"
1050 PRINT " *                                                 *"
1060 PRINT " ***************************************************"
1070 INPUT "                              Choose 1,2 OR 3 ";CHOICES
1080 IF CHOICES="1" THEN GOTO 580
1090 IF CHOICES="2" THEN GOTO 100
1100 IF CHOICES="3" THEN GOTO 1790
1110 GOTO 990
1120 REM Move matrix into a tempory matrix
1130 DIM TM(N,Q)
1140 REM The temporary matrix
1150 FOR I = 1 TO N
1160 FOR J = 1 TO Q
1170 TM(I,J) = M(I,J)
1180 NEXT J
1190 NEXT I
1200 RETURN
1210 REM Eliminate factors
1220 FOR L = N TO 2 STEP -1
1230 GOSUB 1460
1240 REM Sort equations :- highest first factor order
1250 GOSUB 1590
1260 REM Dimension smaller matrix
1270 GOSUB 1630
1280 REM Calculate new lines into the smaller matrix
1290 GOSUB 1710
1300 REM Move smaller matrix into a redimensioned temporary matrix
1310 NEXT L
1320 RETURN
1330 REM DEtermine Xn's value
1340 RESULT(F) = TM(1,2)/TM(1,1)
1350 RETURN
1360 REM Set up next factor to be eliminated
1370 FOR I = 1 TO N
1380 FOR J = 2 TO N
1390 LET TEMP = M(I,J-1)
1400 M(I,J-1) = M(I,J)
1410 M(I,J) = TEMP
1420 NEXT J
1430 NEXT I
1440 REM Note that the constants in column N+1 are not moved
1450 RETURN
1460 REM Sort equations - Highest first
1470 FLAG = 0
1480 FOR I = 1 TO (L-1)
1490 IF TM(I,1) >= TM(I+1,1) THEN GOTO 1560
1500 FOR J = 1 TO (L+1)
1510 TEMP = TM(I,J)
1520 TM(I,J) = TM(I+1,J)
1530 TM(I+1,J) = TEMP
1540 NEXT J
1550 LET FLAG = 1
1560 NEXT I
```

54   Systems modelling

```
1570 IF FLAG = 1 THEN GOTO 1460
1580 RETURN
1590 REM Change the size of the smaller matrix
1600 LET Y = L-1
1610 DIM SM(Y,L)
1620 RETURN
1630 REM Calculate new lines
1640 FOR I = 2 TO L
1650 MF=(TM(I,1)/TM(1,1)) * -1
1660 FOR J = 2 TO L+1
1670 SM(I-1,J-1) = MF * TM(1,J) + TM(I,J)
1680 NEXT J
1690 NEXT I
1700 RETURN
1710 REM Move small matrix into a temporary matrix
1720 DIM TM(Y,L)
1730 FOR I = 1 TO Y
1740 FOR J = 1 TO L
1750 TM(I,J) = SM(I,J)
1760 NEXT J
1770 NEXT I
1780 RETURN
1790 END
```

Program 4.3 Solving the equations using Gaussian elimination

4.4   The specification of reliability

It is necessary to specify not only the performance characteristics of an equipment at the early stage of development, but also the reliability and allied characteristics. This is particularly so if the equipment is being developed and/or manufactured under contract, as is the case with defence equipment. It is essential that at this stage a reasonable and realistic figure for reliability is given. For example, an ocean going liner is being supplied with electric generators. What should their reliability be? It is no good picking a figure from the air (say 95%) and leaving the designer and manufacturer to carry on from there. It is better, once the environment and general use has been decided on, to ask about the maintenance facilities available, how many generators there will be on board, how long a journey (the mission) will last, and what sort of availability the user expects from them. When this last question is initially asked, the answer is generally 100%, and it may have to be pointed out that 100% reliability and availability is impossible.

Let us suppose that the user has settled for a bank of four generators, and is willing to accept 97.5% availability for at least three of them being able to function (i.e. he will accept only three of them working some of the time, but three or all four must be available for at least 97.5% of the time. This gives a measure of effectiveness of the total system.) If we assume that failures are independent, this implies an availability of about 93% for each machine. If the liner is

Worked examples   55

not going to have the facility to repair these machines, then it is necessary to know something about the time at sea and the cost of repair in a foreign port, should the need arise. However if repair facilities are to be carried out, then the repair times must be known. Suppose that it takes on average one hour to collect together the tools, men and equipment needed for a repair (sometimes called the logistic delay), and that the MTTR is 13 hours, making an average downtime of 14 hours. Then the MTBF that is needed to give an availability of 93% is 186 hours.

Other questions can be asked to determine the suitability of this figure. What is the reliability of the generators? To answer that question we must know how long the liner will be at sea. The time between planned maintenance is to be six weeks, we are told. That is 1008 hours. Assuming a constant failure rate for the generators (as they are complex pieces of equipment), if the MTBF is 186 hours, the failure rate is 0.0054, then the reliability after 1008 hours is

$$R = \exp(-0.0054 \times 1008)$$
$$= 0.004$$

i.e. there is almost certain to be a failure in such a time. How many failures are there likely to be? On average there will be $1008/186 = 5.5$ for each of the four generators. There may be more, of course, and Poisson probability paper can be used to determine the risk.

Another problem that should be addressed at this time is that of reliability demonstration testing. The inadvisability of this was mentioned earlier, but if there is to be a demonstration of the achieved reliability at the end of the project, and if this test is to show that some prespecified reliability has been achieved, then it must be possible to demonstrate it realistically. This problem applies particularly to the defence industry, where the user is the forces, but their equipment is developed and produced by private industry.

There is much debate about how best to specify reliability, if indeed it should be specified at all (for contractual purposes, that is). The above is one approach, and to the author's knowledge has not been discussed in the literature before.

WORKED EXAMPLES

(4.1) The ignition system of a car consists of the following components, shown with their reliabilities.

C_1 spark plugs   0.005   (together)
C_2 distributor   0.005   (includes points etc.)

C_3 cables 0.001 (HT and LT)
C_4 battery 0.001
C_5 alternator 0.0005 (includes fan belt)

Assuming the car is running (i.e. has been started), what is the reliability of the ignition system?

The RBD of the system is shown in Figure 4.9.

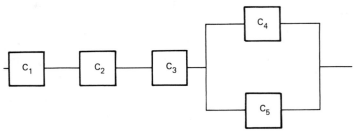

Figure 4.9 Ignition system of a car

The subsystem consisting of C_4 and C_5 must be analysed separately first, in which case,

$$R(C_4 \text{ and } C_5 \text{ together}) = 1 - (1 - 0.001) \times (1 - 0.0005)$$
$$= 0.9999995$$

The system is now a series system, as shown in Figure 4.10, where C_{56} is the system consisting of the battery and alternator together. The reliability of this system is given by

$$R = R_1 \times R_2 \times R_3 \times R_6$$
$$= 0.995 \times 0.995 \times 0.999 \times 0.9999995$$
$$= 0.989$$

Figure 4.10 Ignition system of a car — reduced

(4.2) A four-engined aircraft is able to fly on only one engine once it is airborne, but must have at least one engine functioning on each side during takeoff. The reliabilities are 0.995 during flight but 0.99 during take off. What is the probability that the aircraft will be able to complete its journey?

In order to be able to complete the journey, the aircraft must first take off, and then fly successfully to its destination. These are two separate problems, and have to be analysed independently. First, the take off. The RBD for this part of the journey is shown in Figure 4.11 where P and S stand for port and starboard respectively, and O and I for outer and inner respectively. Using the formulae, the reliability of the port engines together is

$$1 - (1 - 0.99)^2 = 0.9999$$

and similarly for the starboard engines. Then the reliability of the two systems which are now in series is

$$0.9999^2 = 0.9998$$

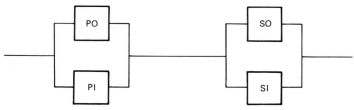

Figure 4.11

The RBD for the engines during the flight of the aircraft is shown in Figure 4.12, and the reliability of this system is

$$1 - (1 - 0.995)^4 = 1 - 6.2 \times 10^{-10}$$

The overall reliability is obtained by multiplying together these two quantities.

This example illustrates that each phase of the mission of a system may have to be analysed separately. Care must be taken to consider every phase when doing this analysis.

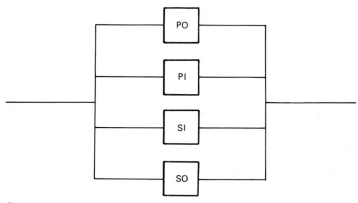

Figure 4.12

(4.3) A manufacturing plant consists of a production process and a tank of dangerous fluid, connected by a pipe. In the pipe is an electronically controlled valve to shut off the flow should it be necessary because of fire or some other hazard. The reliability of the electronic control is 99.5%, i.e. there is a probability of 0.005 that the valve will not close when required to do so. This is not considered sufficiently safe, so a second valve with its own electronics was put in the pipe as shown in Figure 4.13. Although the two valves are in series

Figure 4.13 Pipe-valve system

in the pipe, this is a redundant system, because if one of the valves fails to close the other can take over. The reliability of the two valves together is:

$$1 - (1 - 0.995)^2 = 0.999975$$

The RBD of this system is shown in Figure 4.14.

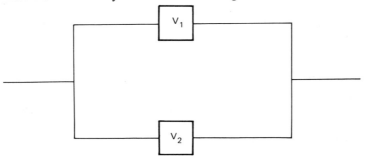

Figure 4.14 RBD of two-valve system

This example illustrates the fact that when constructing RBDs the series redundant structure of a system need not be related to the geometric structure of the system itself. But that is not the end of the story. A safety system like the one described serves two purposes, and both need to be considered. Not only must the valves close when there is a hazard, they must stay open when there is no hazard. (Similarly, a fire alarm serves two purposes, to make a noise when there is a fire, and to keep quiet when there is not.) The RBD for the second function is shown in Figure 4.15, and in this case the system is a series one. So if the reliability when considering this failure mode is 0.997, it has been reduced to

$$0.997^2 = 0.994$$

Figure 4.15 RBD of two-valve system, production viewpoint

The first failure mode is the one considered by the safety manager, and the second that considered by the production manager. The further lesson to be learnt from this example is that a system may well serve several functions, or have several failure modes. They may well need to be analysed separately, and have very different RBDs.

One solution to the problem of improving the reliability for both modes, is to have a 'one-out-of-two-twice' arrangement. This would be a system of valves pumps as shown in Figure 4.16.

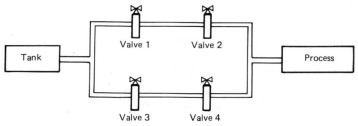

Figure 4.16 Valve systems: 1 out of 2 twice

The old arrangement was a 'one-out-of-two' system. Now there are two of them and they both must function properly if the system is to be safe. The RBD (from the safety point of view) is shown in Figure 4.17.

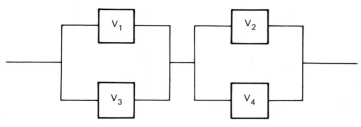

Figure 4.17 RBD of four-valve system, safety view

The reliability is now

$$0.999975^2 = 0.99995$$

This is still extremely high!

From the production point of view, the RBD is that shown in Figure 4.18, and the reliability is:

$$1 - (1 - 0.997^2)^2 = 0.99996$$

which is a great improvement.

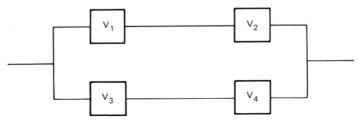

Figure 4.18 RBD of four-valve system, production view

(4.4) Consider the system whose RBD is shown in Figure 4.19.

Suppose that $R_a = 0.98$ and $R_b = 0.96$. It is necessary to improve the reliability of the system by using redundancy; and there is a choice

Figure 4.19

of the system shown in Figure 4.20 (redundancy at system level) or the one shown in Figure 4.21 (redundancy at component level).

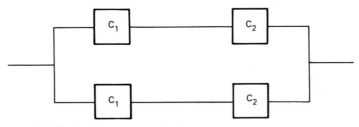

Figure 4.20 Redundancy at system level

For the original system, the reliability is

$$0.98 \times 0.96 = 0.9408$$

For the first system with redundancy the reliability is

$$1 - (1 \times 0.9408)^2 = 0.9965$$

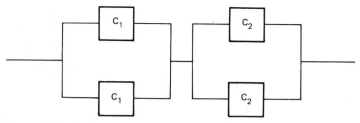

Figure 4.21 Redundancy at component level

For the second system with redundancy the reliability is
$$(1 - (1 - 0.98)^2) \times (1 - (1 - 0.96)^2) = 0.9996 \times 0.9984$$
$$= 0.9980$$
which is greater. This is a special case of the general result that it is better to build in redundancy at component level than at system level.

(4.5) A lone yachtsman, travelling single-handed non-stop around the world has to carry spares on a trip of about 30 000 sea miles. He believes that a particular component of his self-steering gear has a constant failure rate with MDTF (mean distance to failure) of 10 000 sea miles). If he carries two spares (making three components carried in all), what is the probability that he will run out of that particular component?

Solution. As he is carrying three components, two spares and one in use, the curve on the paper labelled three must be used.
$$a = 30\,000/10\,000$$
$$= 3$$
The value $T = 3$ is labelled $A - A'$ on Figure 4.22, and the value of P on curve 3 corresponding to this
$$P = 0.57$$
is labelled $B - B'$.

So he has a 57% chance of running out of this particular component.

This is unacceptable, so how many spares should be carried to ensure a probability of at least 95% that he will not run out of spares?
The value of P corresponding to a reliability of 95% is
$$P = 0.05$$
and the value of a is 3.

The point corresponding to $P = 0.05$ and $a = 3$ lies between the curves 6 and 7, so he must carry seven components to meet the requirement. As this includes the one in use, six spares must be carried. This is shown in Figure 4.23.

This is still unacceptable as space is at a premium. He decides he can carry three spares. What must their failure rate be to ensure a reliability of at least 95%?

The value of $P = 0.05$ intersects curve 4 (3 spares and 1 in use) at the value

$$a = 1.3$$
$$= 30\,000/\text{MDTF}$$

So

$$\lambda = 1/\text{MDTF}$$
$$= 1.3/30\,000$$
$$= 0.000043 \text{ failures per sea mile}$$
$$\text{MDTF} = 23\,000 \text{ sea miles}$$

PROBLEMS

(4.1) An alternative approach to the safety valve problem discussed in Example 3 above is to arrange the valves in the pipe in the arrangement shown in Figure 4.24. Using the same values of the two reliabilities as were used in Example 3, calculate the safety and production reliabilities of the system.

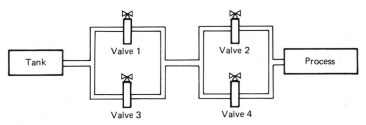

Figure 4.24 Alternative pipe-valve arrangement

(4.2) Calculate the overall reliability of the system with RBD shown in Figure 4.25 on page 65. The reliabilities of the individual components are written next to them in the diagram. Calculate the reliabilities if redundancy is built in (a) at system level, and (b) at component level for each component.

Worked examples 63

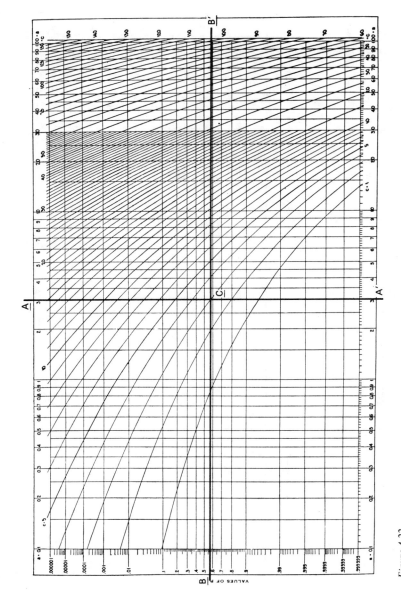

Figure 4.22

64 Systems modelling

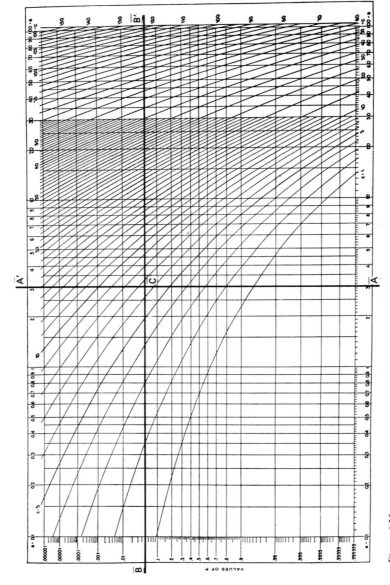

Figure 4.23

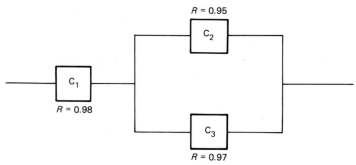

Figure 4.25

(4.3) The yachtsman described above finally decides to carry four spares with a MDTF of 15 000 sea miles. What are his chances of completing his voyage without running out of this component? After completing half of his journey, he has had one failure. What is the probability that he will successfully complete the reaminder of the voyage?
(*Solution:* 95%, 98%.)

(4.4) An arctic expedition expects to be away from home for 18 months. They require a 99% probability of not running out of a particular subassembly of their radio that has a MTTF of four months. Assuming CFR, how many spares should they carry to achieve this objectivejective.
(*Solution:* 10.)

(4.5) A space mission is to last eight days. An essential part of the equipment is replaceable, and the mission can carry one spare. The overall reliability required is 99.5% What should the failure rate of the equipment be if this target is to be met? Having established the failure rate, what would the reliability be if they were to carry no spares?
(*Solution:* 80 days, 92%.)

(4.6) A simple network system consists of a computer and two terminals. It can be in any one of four states. These are, completely functioning, one terminal operational, neither terminal operational, and the computer itself failed (irrespective of the state of the

66   Systems modelling

terminals). The transition matrix for the system is

	S1	S2	S3	S4
S1	0.8	0.25	0.0	0.4
S2	0.15	0.65	0.4	0.15
S3	0.0	0.05	0.55	0.15
S4	0.05	0.05	0.05	0.3

Analyse this system.

(4.7) The above system would actually have eight states, when considering all the possible combinations of functioning and non-functioning for each of the three subsystems. If the failure rate of the terminals is 0.15, and that of the computer is 0.05, and the repair rates are 0.4 and 0.25 respectively, and if it is assumed only one item can be repaired at a time, draw the transition diagram and write down the transition matrix. You should also make the computer repairs have priority over the terminal repairs, i.e. if the computer fails, then work on restoring it will start immediately, even if this means stopping work on a terminal that has already failed.

Most texts on reliability deal with RBDs and the problem of modelling reliability of large complex systems. For a deeper, more mathematical discussion of modelling, Barlow (Reference 2 in Chapter 2), gives a very good account.

Chapter 5

Predicting reliability during the design stage

It is often important that early in the design stage of a piece of equipment there is some attempt made to predict the eventual reliability. Of necessity, this cannot be a precise science, but it is often important that some indication is made of the potential of a design to meet the requirement. Predictions can also be a way in which senior management can make themselves visible to the designer. This high profile approach is one way in which management can communicate their enthusiasm for reliability to the design team. This can be particularly important if the equipment is being developed under contract, and designers and project managers cannot meet frequently.

5.1 Parts count and parts stress

These two prediction techniques are discussed together, as they are very similar, and may well use the same or at least similar data bases.

In a parts count analysis the failure rates of the components (counting multiplicities) are simply summed to give a failure rate of the system or subsystem as a whole. It can be done very early in the design, before any hardware exists. It is not at all accurate on an absolute scale, and there is some debate on its effectiveness, but it is relatively inexpensive and simple, and is very useful for comparing different designs. The effects of redundancy are not taken into account, and as a result its use may be restricted to subsystems with little or no redundancy. Alternatively, the result may be considered as a defect (needing maintenance) rate rather than a failure rate. A more detailed analysis can be done if the design is sufficiently advanced, or if degrees of redundancy are being compared, but if the reliability structure of the item under consideration is very complex, a fault tree analysis may be more suitable.

Parts stress analysis is very similar to parts count, except that the working environment of the components is taken into consideration, and the failure rates are suitably adjusted. The temperature and other stresses of components may be brought into the equation, as well as

68   Predicting reliability during the design stage

any derating of electronic and electrical components. The manufacturing quality of items can also be brought into the equation. This technique is most suitable for electronic and electrical components, as the behaviour of mechanical and other items is much less predictable from theoretical considerations. Much the same comments apply here as with parts count, except that it must be done later in the design, when some of the hardware has been put together, if only to allow the designer to measure the stresses on the components, due, for example, to the geometry of the design.

This very simple technique is discussed in most reliability texts. The source of data most often used, at least in the defence industry, is MIL-HDBK 217 (Reference 20 in Chapter 2).

There is no program for this section. Software able to perform the simple arithmetic needed for a parts count and parts stress is little more than a data base, which is not a suitable type of program for inclusion in a text of this nature (MIL-HDBK 217 is bigger than this book).

5.2   Failure mode effect and criticality analysis

In this section we consider not only the failure of a component or subassembly, but the possible ways it may fail (the failure mode), and the effects different failure modes may have on the system (the severity level). It must be said at this point that failure modes effects and criticality analysis (FMECA) is not a precision tool, and is regarded by some practitioners as a bit of a blunt instrument. However, it does give designers, project leaders and managers an insight into the reliability structure of a complex system, and underlines the comparative strengths and weaknesses of the various subassemblies of the system. It is a bottom-up approach, in that each component, or subassembly if it is not considered desirable to do the analysis in more detail, is examined for the effect of its failure would have on the system as a whole, rather than the top-down approach unlike, for example, a fault tree analysis. An FMECA is not suitable for a system with a lot of redundancy, as the failures are considered in isolation. Again, a fault tree analysis may be more suitable for this situation.

There are two possible ways of performing an FMECA: quantitative and qualitative. They are not so very different, and both will be described, starting with the quantitative. By assigning a numerical value to the severity of each failure mode, and then combining these with the failure rate of the component in question, we obtain the criticality of that failure mode, and by adding the criticalities together, a criticality for each component is obtained. The sum of the criticalities of each component in an assembly or subassembly gives a criticality of each assembly, and can thus be used to indicate possible areas for future design action. A high criticality value indicates redesign for example, or maybe a change to more reliable, and hence more expensive, components, or special attention during routine maintenance. This is best illustrated by an example, and Figure 5.1 shows a drum assembly of an imaginary washing

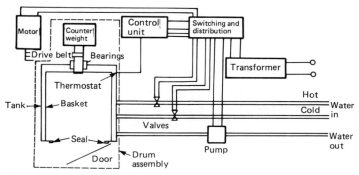

Figure 5.1 Schematic diagram of a hypothetical washing machine

machine. The FMECAs shown in Tables 5.1 and 5.2 have the following entries:

1–3. The component under discussion, a reference number and code number. This is self-explanatory.
4. The components function (or functions, if there is more than one).
5. The failure modes of the components i.e. the different ways in which it might fail.
6. The failure mode ratio. This is the proportion of failures of the component in question that turn out to be that particular failure mode. The sum of the failure mode ratios for each item should be one. In practice this is not usually the case, as most authors of FMECAs are unwilling to put their hands on their hearts and claim that they have considered all possible failure modes. The excuse given is that any that are left out are either so rare, or so non-critical, or both, as to be not worth bothering about. However, if the sum of the failure mode ratio is much less than 1, say 0.8 or less, this would suggest that something is not quite right. The failure mode ratio is denoted α.

FAILURE MODES AND EFFECTS ANALYSIS – WORKSHEET

SYSTEM Automatic washing machine   SYSTEM Water   ASSEMBLY Basket

Item	Ref. No.	Code No.	Function	Failure mode	Failure frequency (α)	Failure $(10^6\,hrs)$ (λ)	Failure effect Immediate level*	Failure effect Next level†	Symptoms	Severity level (S)	Criticality (C) $(=\alpha\lambda S)$	Remarks and total criticality for each item
1	2	3	4	5	6	7	8	9	10	11	12	13
Basket	B	B/20/36	Holds clothes	Warp	0.5		May impede rotation		Noise	0.8	0.16	
				Crack	0.5	0.4	Impede rotation and may damage clothes			1.0	0.20	0.36
Shaft	S	S/18/2	Rotates basket	Break	0.1					1.0	0.06	
				Wear	0.7	0.6	Total failure	Could damage machine	Noise	0.7	0.29	0.43
				Deform	0.2					0.7	0.08	
Bearings	B4	B/2/2		Loss of lubrication	0.45	0.7	Slow action	Clothes not properly washed	Noise	0.6	0.19	
				Contamination	0.3		Impede rotation	As above		0.6	0.13	
				Misalignment	0.05			Could lead to catastrophic leak		1.0	0.04	
				Brinelling	0.05					0.2	0.01	
				Corrosion	0.15			Possible leak and/or damaged shaft		0.8	0.08	0.44
Drive belt	Db	Db/502 (a)		Break	0.15	0.9	No rotation	No functioning		1.0	0.14	
				Wear	0.8		Slow rotation	Clothes not properly washed		0.5	0.36	0.50
Counter-balance	Cb			Crumble	0.5	0.1	Eventually damage shaft and bearings and			0.3	0.015	
				Fall off	0.5		more catastrophic. Could cause extensive damage			1.0	0.05	0.065

* Immediate level for this analysis   *Assembly*

† Next level for this analysis   *Subsystem*

Total criticality   1.80

Table 5.1 FMECA of the drum assembly of an automatic washing machine

FAILURE MODES AND EFFECTS ANALYSIS – WORKSHEET

SYSTEM Automatic washing machine SUBSYSTEM Water ASSEMBLY Tank

Item	Ref. No.	Code No.	Function	Failure mode	Failure mode frequency (α)	Failure rate (10^6 hrs) (λ)	Failure effect — Immediate level*	Failure effect — Next level†	Symptoms	Severity level (S)	Criticality (C) ($=\alpha\lambda S$)	Remarks and total criticality for each item
1	2	3	4	5	6	7	8	9	10	11	12	13
Tank	T	TA1/4	Hold water	Warp Leak Burst	0.3 0.4 0.25	0.05	Possibly foul moving parts Catastrophic		Noise Wet floor	0.7 0.7 1.0	0.01 0.014 0.013	Total 0.037
Hot water pipe	P1	Cu/1/2		Warp Leak Burst	0.5 0.4 0.05	0.1	May impede flow Catastrophic		Fills slowly Wet floor	0.1 0.7 0.1	0.005 0.028 0.005	Total 0.038
Cold water pipe	F2	Cu/1/2		Warp Leak Burst	0.3 0.6 0.05	0.05	As above		Fills slowly Wet floor	0.1 0.7 1.0	0.002 0.021 0.002	Total 0.025
Hot water valve	V1	V124	Control flow of hot water	Leak Burst Seizes shut Seizes open	0.3 0.05 0.5 0.1	1.0	Leak Burst No hot water Overflow	Cold wash	Wet floor As 8	0.7 1.0 0.4 1.0	0.21 0.05 0.2 0.1	0.56
Cold water valve	V2	V124	Controls cold water	Leak Burst Seizes shut Seizes open	0.3 0.03 0.5 0.1	1.0	Catastrophic No cold water overflow			0.7 1.0 1.0 1.0	0.21 0.05 0.5 0.1	No cold water Could ruin clothes Total 0.86
Hot and cold connectors (two in series)	C1 C2	CCu/1/2 Ccu/1/2		Leak Burst	0.6 0.3	0.1	Catastrophic		Wet floor	0.7 1.0	0.041 0.03	0.077 each 0.144 together
Drain pipe to pump	P5	Cu/1/2	Carries water away	Leak Burst Warp	0.5 0.3 0.1	0.1	Catastrophic May impede flow		Empties slowly	0.2 0.1 0.1	0.035 0.03 0.001	0.066
Drain connection	C5	Ccu/1/2		Leak Burst	0.9 0.1	1.0	Catastrophic		Wet floor	0.7 1.0	0.63 0.1	0.73
Seal	S2	S/R/15		Leak	1.0	0.1	Catastrophic			0.6	0.16	0.16
Door	D	D/128		Warp Break	0.85 0.1	0.1	May leak Flood			0.7 1.0	0.06 0.01	0.07
*Immediate level for this analysis		*Assembly*									Total criticality	2.69
†Next level for this analysis		*Sub-system*										

Table 5.2 FMECA of the tank and pipework of an automatic washing machine

72 Predicting reliability during the design stage

7. The failure rate of the component in question, denoted by λ as usual. Most FMECAs assume a constant failure rate, and hence put the components in the useful life part of their bathtub curve. $\lambda\alpha$ will give the failure rate for the failure mode with failure mode ratio α.

8–9. The effects of the failure mode on the immediate level (subassembly) and the next level (assembly). For a piece of equipment as simple as a washing machine this is quite sufficient, but for a complex item of equipment, such as an aircraft, the effect on several different levels may have to be discussed. In particular, the effect on the total equipment, or 'the mission' should be discussed here.

10. Symptoms. These should be different from, or add to, anything in columns 8 and 9 if at all possible.

11. The severity level. The simplest definition is that the severity level is the proportion of times that a particular failure mode causes the system to fail, but this has its difficulties when degraded systems are considered, or possibly in gathering sufficient data, or when possible loss of life or severe injury or financial loss may be caused by particular failure modes, for example. If the equipment has a 'mission', then one can talk of loss of the mission, but most items of equipment do not have a single, uniquely defined mission. It may be that the assigning of the severity level has to be done in a subjective manner, 1 being assigned to a failure mode the effects of which are as bad as can be, while 0 is assigned to failure modes that do not matter at all. There is no objection to such a course of action, provided that the rules for deciding the values are carefully considered and decided beforehand, and the severity levels themselves are, broadly speaking, agreed by all concerned, and are not the work of one man. It may be that the qualitative approach, as described below, would be better. The severity level is denoted S.

12. Criticality. The criticality of each failure mode is given by

$$C = \lambda\alpha S$$

and the total criticality of each component is given by

$$C = \sum \lambda\alpha S$$

summed over all the failure modes of the component, while the criticality of each subassembly is given by $C = \sum C$, summed over each component of the subassembly.

Failure mode effect and criticality analysis   73

In doing these FMECAs the following guide was used in determining the values of the severity levels:

Failure	Severity level
Unable to operate, floods, damage to clothes	1.0
Leak (not flood)	0.7
Impaired operation (i.e. fills slowly)	0.2–0.6

The reader will notice that although the failure rates of the two subassemblies are very similar, the criticality of the drum assembly is about 50% greater than that of the tank. The implication of this is that although the two assemblies will suffer about the same number of breakdowns in a given period, the tank and associated components will give fewer critical breakdowns leading to flooding, total failure or damaged clothes. If it is considered that more design effort should be put in, then resources applied to the drum assembly will be more cost-effective in improving the reliability of the machine. Similar comments can be made about the components. It will be observed that the components can be divided into two classes, with the basket, shaft, bearings, drive and the valve all having criticalities greater than 0.45, while all the other components have criticalities less than half that value. These then, are the components that require most attention in order to improve the reliability.

The qualitative approach is similar, except that the severity level is put into one of four categories, 1 to 4, 1 being the most severe. Def. Stan. 00-41 (Reference 18 in Chapter 2) suggests the following classification system:

- Category 1   *Catastrophic.*   A failure which may cause death or system loss.
- Category 2   *Critical.*   A failure which may cause severe injury or damage, and which would result in mission failure.
- Category 3   *Major.*   A failure which may cause minor injury or damage, and which might delay or degrade the mission.
- Category 4   *Minor.*   A failure which would not cause injury or damage, but which would require unscheduled maintenance.

The defence standard was written for more complex and potentially dangerous equipment than our washing machine, but Tables 5.3 and 5.4 show a FMECA done on the same drum assembly and tank and pipework as Tables 5.1 and 5.2, but this time the qualitative approach was used. The categories used were the ones quoted above from Def. Stan. 00-41, but with the appropriate interpretation for a washing machine. The columns in Tables 5.3 and

FAILURE MODES AND EFFECTS ANALYSIS – WORKSHEET

SYSTEM   Automatic washing machine   SUBSYSTEM   Water   ASSEMBLY   Basket

Item	Ref. No.	Code No.	Function	Failure mode	Failure rate	Failure effect		Symptoms	Severity level	Remarks
						Immedite level*	Next level†			
1	2	3	4	5	6a	8	9	10	11a	13
Basket	B	B/20/36	Holds clothes	Warp	0.2	May impede rotation		Noise	2	
				Crack	0.2	Impede rotation and may damage clothes			1	
Shaft	S	S/18/2	Rotates basket	Break	0.06	Total failure	Could damage machine	Noise	1	
				Wear	0.42				3	
				Deform	0.12				3	
Bearings	B4	B/2/2		Loss of lubrication	0.315	Slow action	Clothes not properly washed	Noise	2	
				Contamination	0.21	impedes rotation	As above		3	
				Misalignment	0.035	Could lead to catastrophic leak			2	
				Brinelling	0.035	Possible leak and/or damaged shaft			2	
				Corrosion	0.105					
Drive belt	Db	Db/502 (a)		Break	0.135	No rotation	No functioning		1	
				Wear	0.72	Slow action	Clothes not properly washed		3	
Counter-balance	Cb			Crumble	0.05	Eventually damage shaft and bearings and			2	
				Fall off	0.05	more catastrophic. Could cause extensive damage			1	

* Immediate level for this analysis   *Assembly*
† Next level for this analysis   *Sub-assembly*

Table 5.3 FMECA of the drum assembly of an automatic washing machine

SYSTEM   Automatic washing machine   SUBSYSTEM   Water   ASSEMBLY   Tank

Item	Ref. No.	Code No.	Function	Failure mode	Failure rate	Failure effect Immedite level*	Next level†	Symptoms	Severity level	Remarks
1	2	3	4	5	6a	8	9	10	11a	13
Tank	T	TA1/4	Hold water	Warp	0.015	Possibly foul moving parts		Noise	2	
				Leak	0.02			Wet floor	2	
				Burst	0.0125	Catastrophic			1	
Hot water pipe	P1	Cu/1/2		Warp	0.05	May impede flow		Fills slowly	4	
				Leak	0.04			Wet floor	2	
				Burst	0.005	Catastrophic			1	
Cold water pipe	P2	Cu/1/2		Warp	0.015	As above		Fills slowly	4	
				Leak	0.03			Wet floor	2	
				Burst	0.0025				1	
Hot water valve	V1	V124	Control flow of hot water	Leak	0.3			Wet floor	2	
				Burst	0.05	Catastrophic			1	
				Seizes shut	0.5	No hot water	Cold wash	As 8	3	
				Seizes open	0.1	Overflow			1	
Cold water valves	V2	V124	Controls cold water	Leak	0.3				2	
				Burst	0.03	Catastrophic			1	No cold water
				Seizes shut	0.5	No cold water			1	Could ruin clothes
				Seizes open	0.1	Overflow			1	
Hot and cold connectors (two in series)	C1 C2	CCu/1/2 CCu/1/2		Leak Burst	0.06 0.03	Catastrophic		Wet floor	2 1	
Drain pipe to pump	P5	Cu/1/2	Carries water away	Leak	0.05				2	
				Burst	0.03	Catastrophic		Empties slowly	1	
				Warp	0.01	May impede flow			4	
Drain connection	C5	CCu/1/2		Leak	0.9			Wet floor	2	
				Burst	0.1	Catastrophic			1	
Seal	S2	S/R/15		Leak	0.1				2	
Door	D	D/128		Warp	0.08	May leak			3	
				Break	0.01	Flood			1	

* Immediate level for this analysis   *Assembly*
† Next level for this analysis   *Sub-system*

Table 5.4 FMECA of the tank and pipework of an automatic washing machine

5.4 are similar to those in Tables 5.1 and 5.2, with the following exceptions.

Column 6a. Failure rate. This is the failure rate of the failure mode under consideration, and so is the product of the item failure rate and the failure mode ratio. This is an alternative way of presenting this particular item of data, and from the point of view of the author of the FMECA, conceals the fact that the sum of the failure mode ratios for a given item may not be one.

Column 11a. The severity level. This is defined above, and takes the one of the values 1 to 4. There is now no column 12.

5.3 Criticality matrices

The results of the last FMECA are not so easily compared as in the first case, and one way round this problem is by use of criticality matrices. Two examples are shown in Figures 5.2 and 5.3, for the two subassemblies already considered. They are no more than histograms of the occurrences of the severities 1 to 4, but instead of the frequency being shown on the vertical axis, the total of the failure rates of failure modes with the severity concerned. Thus it can be seen that the 'lump' on the severities 1 and 2 in Figure 5.2 indicates that this subassembly calls for more design action. On the grounds that a picture is worth a thousand words, this pictorial way of presenting the data gives almost as much information as the quantitative conclusion of the first FMECA.

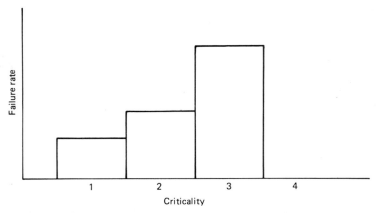

Figure 5.2 Criticality matrix for the tank assembly of a washing machine

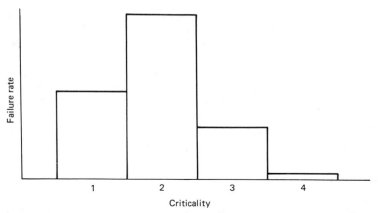

Figure 5.3 Criticality matrix for the basket assembly of a washing machine

FMECA is a very effective tool for detecting design weaknesses, directing the designers (and others) attention to them, and for verifying designs. It can be an expensive activity, as it is very labour-intensive, and so it is important that it is done properly. There is a tendency, particularly when equipment is being procured, for example with the defence industry, for the design authority to hand the problem over to a consultant under a contract (making it more expensive, of course). Since talking with engineers in private industry, the author is led more and more to the conclusion that the people to perform the analysis are the designers, with input from anybody with experience of the equipment such as representatives from the user, the maintenance authority and the production department. An input from an independent source, a 'fresh mind', can also be invaluable. The problem is that somebody new looking at the design has to familiarize themself with it. Going up this learning curve takes time and hence costs money, putting up the cost again.

There is the possibility of further data to be included in the FMECA report if it is desired, for example, recommendations for maintenance, examination and care by the user. Recommendation for production and manufacture can be added perhaps, with respect to quality control or quality assurance of the finished product or some of its components. The list is as long as is required.

FMEA and FMECA are not the delicate tools that some statisticians and engineers would like, and for that reason they are very often omitted from textbooks. However, O'Connor (Reference 12 in Chapter 2) and Carter (Reference 4 in Chapter 2) both give adequate coverage of this most useful design tool. Def. Stan. 00-41 (Reference 18 in Chapter 2), BS 5760 (Reference 15 in Chapter 2) and MIL-STD 1629 (Reference 21 in Chapter 2) are also well worth reading on this topic.

It was felt by the author that it was not worth putting in a program to perform a FMECA. Similar comments apply here as those already made about parts count and parts stress, namely, that such a program should be the front end to a large database. Omitting this facility, as well as file handling facilities to deal with the results, would leave a very trivial program.

5.4 Fault tree analysis

It may be that we wish to analyse all the possible combinations of faults that may give rise to no hot water in our washing machine. This could be due to the heater in the machine itself not working properly, or no hot water arriving through the appropriate pipe. These events will also require analysing, and this can be done with the aid of the tree-like structure shown in Figure 5.4.

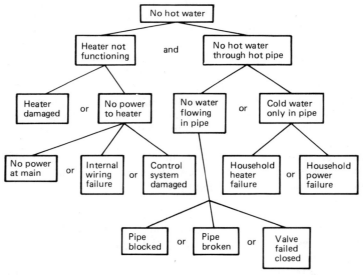

Figure 5.4 Informal fault tree analysis of washing machine water heating system

This arrangement can be laid out in a more formal way as illustrated in Figure 5.5. The pattern of this is exactly the same as that of Figure 5.4, but now symbols are used instead of the words 'or' and

'and'. These symbols are called gates, and are illustrated below. The situations described in the boxes are called events, and each gate has an output event and a number of input events. The output events are drawn above the gates in the tree, and the input events below, and the outputs are the consequence of all (in the case of an 'and' gate) or just one (in the case of an 'or' gate) of the input events. There is just one event that is an output but is not an input to any gate, called the top event. In our example it is 'no hot water in the washing machine'. There are a number of events that are inputs but not outputs. These are called basic events. Some of these are of such a basic nature that they are not worth further analysis, while others may be analysed at some other stage of development. The former are called primary basic events, while the latter are called secondary basic events, and are drawn in circles and diamonds respectively. The non-basic events, i.e. those that we do analyse, are drawn in rectangles. All this symbolism is shown in Figure 5.6.

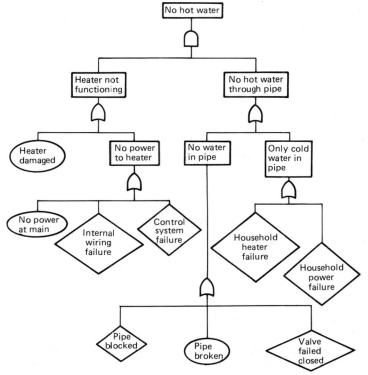

Figure 5.5 FTA of the heating system of a washing machine

80 Predicting reliability during the design stage

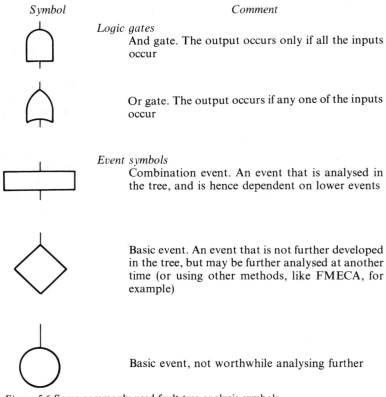

Figure 5.6 Some commonly used fault tree analysis symbols

5.5 Minimum cut sets

Consider the following combinations of failures.

1. heater failed and no power to main;
2. heater failed and valve failed closed;
3. heater failed and no power to main and valve failed closed.

Combination 1, although a multiple failure, will not fail the system, but the other two combinations will. For this reason combinations 2 and 3 are called cut sets (of failures). Now consider combination 2. If any of the two failures are repaired, the system will come back up, i.e. will start to function again, while if the combination described in 3 occurs, the only single repair that will bring the system up again is to treat the failed valve. Repairing the heater without restoring power from the main will leave the system in the failed state. Combination 2 is called a minimum cut set, or min. cut. More formally:

1. A cut set is a combination of component (or subsystem or whatever) failures that ensures system failure.
2. A min. cut is a cut set that is minimal, i.e. repairing any of the failures listed in the cut set is sufficient to bring the system back up.

It is naturally of interest to know the min. cuts of a system. While it is not hard to write down the min. cuts of the simple system described above, a more complex system, especially one involving redundancy, or multiple use of some units, in which case a device may appear at more than one place in the fault tree, would be more difficult. An algorithm will now be described here that generates all the min. cuts of a system, provided the fault tree has been properly drawn. This algorithm will be illustrated first using the example above, and then using two more examples of increasing complexity. It is suggested that the reader tries analysing the two examples before reading the explanations of the text.

First write down the top event:

No hot water

This event is the output to an and gate, so replace it by the gate's inputs written horizontally:

Heater not functioning, no hot water through pipe

First analyse the heater failure. This is the output to an or gate, so replace it by the inputs to this gate. This time write them vertically, repeating the other failure on each line:

Heater damaged,   no hot water in pipe
No power to heater,   no hot water in pipe

Each of the lines is a cut set, but neither of them are made up of basic events. At the end of the analysis, as we replace each non-basic event by the inputs to the appropriate gate, each line of events will be a cut set, and eventually only basic events will appear. Analysing the 'No hot water in pipe' failure, the output to an or gate, and replacing it by the inputs to the gate, we get

Heater damaged,   no water in pipe
Heater damaged,   only cold water in pipe
No power to heater,   no water in pipe
No power to heater,   only cold water in pipe

82 Predicting reliability during the design stage

This procedure is carried on until there are only basic events:

Heater damaged,   pipe blocked
Heater damaged,   pipe broken
Heater damaged,   valve failed closed
Heater damaged,   household heater failure
Heater damaged,   household power failure

No power at main,   pipe blocked
No power at main,   pipe broken
No power at main,   valve failed closed
No power at main,   household heater failure
No power at main,   household power failure

Internal wiring failure, pipe blocked
Internal wiring failure, pipe broken
Internal wiring failure, valve failed closed
Internal wiring failure, household heater failure
Internal wiring failure, household power failure

Control system failure, pipe blocked
Control system failure, pipe broken
Control system failure, valve failed closed
Control system failure, household heater failure
Control system failure, household power failure

All the min. cuts are generated by this procedure, and in this simple case, all the cut sets generated are min. cuts.

Now consider the cooling system illustrated in Figure 5.7. It consists of two pumps, P1 and P2, two sensors, S1 and S2, and two generators, G1 and G2. Each generator serves one pump and

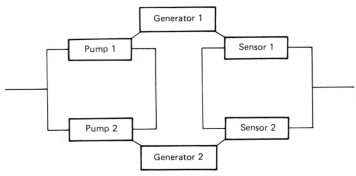

Figure 5.7 Cooling system

supplies power to one sensor. Furthermore, the system will continue to function satisfactorily if only one pump and only one sensor is functioning. The fault tree for this system is shown in Figure 5.8. Notice that each generator appears in two places in the tree.

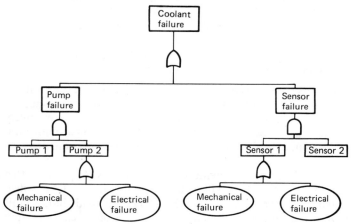

Figure 5.8 FTA of cooling system

The symbols denoting the appropriate failures are shown in the appropriate places in the tree. The analysis proceeds as follows. Coolant failure is the top event:

Coolant failure

Coolant failure is the output to an or gate with inputs pump failure and sensor failure. Replace coolant failure by pump failure and sensor failure written vertically:

Pump failure
Sensor failure

Pump failure is the output to an and gate with inputs Pump 1 and Pump 2. Replace pump failure with these inputs written horizontally:

Pump 1, Pump 2
Sensor failure

(Notice that sensor failure still only appears once, unlike the case when an or gate is analysed.) Sensor failure is the output to an and gate, with inputs Sensor 1 and Sensor 2. Replace sensor failure with these inputs written horizontally:

Pump 1,   Pump 2
Sensor 1,   Sensor 2

84 Predicting reliability during the design stage

Continuing in this fashion, we eventually get:

Pump 1 (mechanical),	Pump 2 (mechanical)
Pump 1 (mechanical),	Generator 2
Generator 1,	Pump 2 (mechanical)
Generator 2,	Generator 2
Sensor 1 (mechanical),	Sensor 2 (mechanical)
Sensor 1 (mechanical),	Generator 2
Generator 2,	Sensor 2 (mechanical)
Generator 1,	Generator 2

All the min. cuts are in this list but in this case, because of the multiple use of the generators, one of the cut sets appears more than once. In the use of this algorithm for systems with a complex reliability structure care has to be taken that some of the cut sets are not min. cuts. All the min. cuts are generated, but there may be some dross as well.

FTA is in some senses complementary to FMECA. A system with a complex reliability structure will be better analysed by FTA, while a series system is better served by FMECA. Having said that, a fault tree will only analyse one fault at a time, and it is often safety-related failures that are analysed in this way. Further advantages of FTA are that it is possible to do the analysis with no data, as the list of min. cuts gives insight into the reliability structure, and human activities can appear as events, and hence as events in the min. cuts. This has obvious implications in the ergonomics of a design. The two techniques need not be exclusive, however, and a FTA can take the analysis to a certain point, when FMECA will take over. Consider the example at the start of the section on FTA, analysing the event 'no hot water in the washing machine'. The failure of the components further down the tree will best be served by FMECA rather than FTA in this case.

There is no program for this section, for similar reasons given for the other sections of this chapter. A program would either be too complex for this text, or so trivial as to make it not worthwhile.

As with FMEA, FTA is not always mentioned in reliability texts. However several references in Chapter 2, including O'Connor (Reference 12), Carter (Reference 4) and Barlow (Reference 2) all have a chapter on fault trees. Carter compares and contrasts the use of FTA and FMECA very well. The British Standards mentioned in Chapter 2, BS 5760 (Reference 15) and Def. Stan. 00-41 (Reference 18) also describe this technique.

PROBLEMS

(5.1) Perform a FMECA on a subassembly that is familiar to you. The distributor of a car is a good example. Consider the information that you wish to go in it. If you were a project manager, what information would you consider the most useful? It is probably best to do this exercise in small groups of five or six.

Where does your data come from? Can you justify it? Have you defined criticality and/or severity? Compare and contrast the reports from different groups, and discuss the reasons for any differences.

(5.2) An air conditioning plant for a computer complex consists of five electrically powered pumps, five power sources and a system of pipes, valves and sensors. A critical part of the complex requires ten units of treated air, and it is necessary to analyse the fault: 'less than ten units of air in the critical part of the system'. Two of the pumps can each deliver six units, while the other three can deliver four each. The valve/sensor subsystem that deals with the critical section of the plant can be thought of as a two-out-of-three hot redundant system,

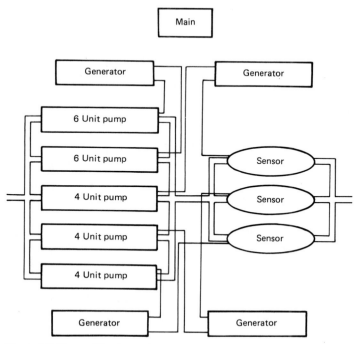

Figure 5.9 Air conditioning system

86 Predicting reliability during the design stage

with each of the three subsystems (VSS) capable of delivering six units. The five power sources consist of the electrical main, which can deliver enough electrical power to drive all the pumps and VSS, while the other four sources are identical diesel generators in a standby mode that can each give enough power to drive either any two pumps, any one pump and one VSS or two VSS. Figure 5.9 shows the pump/VSS layout with one possible configuration for the generators. Using FTA, analyse this configuration. What other configurations are possible? Use FTA to examine possible options to minimize the likelihood of failure.

Notice that no data is given with this problem. It is reasonable to assume that the failure rates of the subsystems are of the same order of magnitude. Comment on this in your solution.

Chapter 6

Estimating reliability

During development the engineer starts to see data generated from his project for the first time. This data can be analysed in a variety of ways, to a variety of ends. A brief discussion of some of the techniques and objectives is included in this chapter.

6.1 Reliability growth

As soon as the prototypes of the equipment being developed during a project are built and tested, failures start to manifest themselves. The disciplined approach is to investigate the cause of each failure, and redesign to try and make sure it will not reoccur (a process known as test, analyse and fix). All too often excuses are made about failures (random failure, non-accountable, not typical, unusual stress levels etc.) and little is done about an individual failure mode until it manifests itself a number of times. Investigating each and every failure is good engineering. In practice it is more usual to have a rule of thumb like 'investigate and design out persistant failures', i.e. those that occur repeatedly. To deal with those that occur twice is good. As a rule of thumb, to allow a failure mode to occur more than three times is being careless.

All that is written in the previous paragraph must also be considered in the system being developed, with an eye on the criticality of the individual failures. For any development programme, if it is properly managed, failures will become less and less frequent, and the times between failures will tend to increase. It is this process of seeing fewer and fewer failures as the system is redesigned to remove the cause that is known as reliability growth.

It is desirable to have a quantitative model of reliability growth, in order to:

1. assess the effectiveness of the development process;
2. estimate how much more development effort is needed to ensure a reliability target is reached, or
3. estimate the final reliability of a product for a given amount of development effort.

88   Estimating reliability

A number of models exist, but we shall only discuss the Duane model here.

Consider the data set shown in Table 6.1 which shows the cumulative times of failure of pump for a washing machine:

Table 6.1

Failure number	Cumulative time of failure (h)	Cumulative MTBF M_c	$\ln(T)$	$\ln(M_c)$
1	103	103	4.6	4.6
2	315	157	5.7	5.1
3	801	267	6.7	5.6
4	1183	296	7.1	5.7
5	1345	269	7.2	5.6
6	2957	493	8.0	6.2
7	3909	558	8.3	6.3
8	5702	713	8.6	6.6
9	7261	807	8.9	6.7
10	8245	824	9.0	6.7

The second column is the cumulative time of the nth failure, where n is the number in the first column, i.e., the fifth failure was observed when a total of 1345 h of testing had been done on all the prototypes together. The third column is the cumulative MTBF, i.e. the number in the second column divided by the number in the first. The last two columns are self-explanatory, and are used to draw Figure 6.1.

From a brief look at the data, it can be seen that there is some effort going in to making the pump design more reliabile, because the failures are becoming less and less frequent as weaknesses are designed out. The problem is to decide whether enough effort has gone into the design and development and when to stop. Figure 6.1 shows the plot of $\log(M_c)$ against $\log(t)$. Duane noticed that this plot approximated very well to a straight line for a large variety of equipment. In practice, the plot is usually made on log–log paper, as in Figure 6.2.

This straight-line fit (which can be done using least squares, if an analytic approach is preferred over the 'fit by eye') gives the relationship

$$\log(M_c) = \alpha \log(t) + \log(A)$$

or

$$M_c = At^\alpha$$

for some α and A.

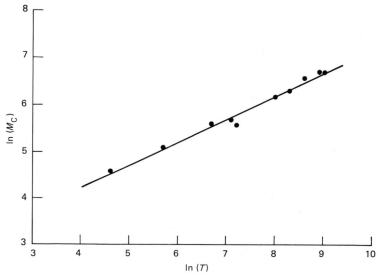

Figure 6.1

α is known as the growth parameter, and lies between zero and one. For our example with the pumps, we have

$$\alpha = 0.5, \quad A = 10$$

so that

$$M_c = 10t^\alpha$$

The larger the value of α the more effective is the development programme, and O'Connor (Reference 12 in Chapter 2), suggests the following guide in interpreting the value of α:

- $\alpha > 0.4$ Reliability has top priority, very effective development programme
- $\alpha = 0.3\text{--}0.4$ Reliability has high priority
- $\alpha = 0.2\text{--}0.3$ Routine attention paid to reliability, important failure modes investigated and analysed
- $\alpha < 0.2$ Reliability has low priority.

The parameter α, then, measures the effectiveness of the programme, and so deals with the first point. It is also desirable to measure the instantaneous reliability. Assuming an instant failure rate, it becomes necessary to know the instantaneous MTBF, M_i.

90 Estimating reliability

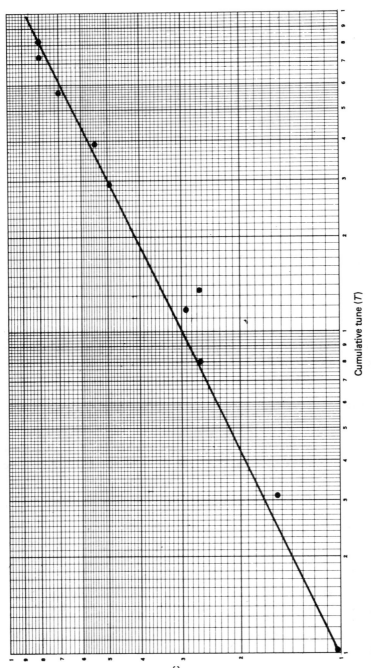

Figure 6.2

That is, if the development were to stop at the last failure time, what would the MTBF of the system be? It is certainly not M_c, but is somewhat higher, because the early design has been modified to reduce the probability of certain failure modes that manifested themsevles early on. Let λ_c, λ_i be the corresponding failure rates, then

$$\lambda_c = 1/M_c, \qquad \lambda_i = 1/M_i$$

so that

$$\lambda_c = n/t$$

where n is the number of failures at time t. This gives

$$n = \lambda_c t$$
$$= t/M_c$$

and

$$M_c = At^\alpha$$

from the Duane plot. So that

$$n = t^{1-\alpha}/A$$

Now λ_i is the failure rate, and so is the derivative of the number of failures with respect to time. We cannot really differentiate the number of failures because it must be a whole number. What we are really differentiating is the number of failures expected, on average, by time t. So put

$$\lambda_i = \mathrm{d}n/\mathrm{d}t$$
$$= (1 - \alpha)t^{1-\alpha}/A$$
$$= (1 - \alpha)/AT$$
$$= (1 - \alpha)\lambda_c$$

or

$$M_i = M_c/(1 - \alpha)$$

or

$$\log(M_i) = \log(M_c) - \log(1 - \alpha)$$

As $1 - \alpha$ is less than 1, M_i is greater than M_c. The last result means that the plot for M_i is a straight line, parallel to the one for M_c, but distance $1/(1 - \alpha)$ above it.

In our example, this gives

$$M_c = 824\ \mathrm{h}$$

92 Estimating reliability

At the last failure (i.e. after 824 h of testing)

$$M_i = M_c/(1 - \alpha)$$
$$= 1648 \text{ h}$$

Suppose we are interested in a cumulative running time of 500 h. The reliability at 500 h (assuming a constant failure rate) is

$$R(500) = \exp(-500/1648)$$
$$= 74\%$$

This is too low. If we require a reliability of 95%, then

$$R(500) = \exp(-500/\text{MTBF})$$
$$= 0.95$$

Taking logs,

$$-500/\text{MTBF} = \ln(0.95)$$
$$\text{MTBF} = 500/0.05$$
$$= 10\,000 \text{ h}$$

This is the instantaneous MTBF, M_i, and

$$M_i = M_c/(1 - \alpha)$$
$$= 2 \times 10t^{1/2}$$
$$= 20t^{1/2}$$
$$t = (10\,000/20)^2$$
$$= 250\,000 \text{ h}$$

This is clearly unrealistic. If development were to continue until a total of 20 000 h testing had taken place, what would the reliability then be?

$$M_i = 20 \times (10\,000)^{1/2}$$
$$= 2828 \text{ h}$$
$$R(500) = \exp(-500/2828)$$
$$= 84\%$$

Program 6.1 finds the straight line using the method of least squares. Once the slope parameter has been calculated, the operator has the option of calculating the MTBF after a stated amount of development time, or the further development time needed to attain a stated MTBF.

Reliability growth   93

```
100 REM
110 REM This routine does reliability growth analysis
120 REM using the Duane model.
130 PRINT ' How many failures';
140 INPUT N
150 DIM T(N),M(N)
160 PRINT ' Input the times at which the failures occurred.'
170 PRINT ' In order-time of first failure';
180 INPUT T(1)
190 M(1)=T(1)
200 FOR I=2 TO N
210 PRINT ' and the time of the ';I;'th failure'
220 INPUT T(I)
230 M(I)=T(I)/I
240 NEXT I
250 REM Calculation of the parameters
260 XSUM,YSUM,XYSUM,X2SUM =0
270 FOR I=1 TO N
280 Y= LOG(M(I))
290 X= LOG(T(I))
300 XSUM=XSUM+X
310 YSUM=YSUM+Y
320 XYSUM=XYSUM+(X*Y)
330 X2SUM=X2SUM+(X*X)
340 NEXT I
350 XMEAN=XSUM/N
360 YMEAN=YSUM/N
370 TOP=(N*XYSUM)-(XSUM*YSUM)
380 BOT=(N*X2SUM)-(XSUM*XSUM)
390 B=TOP/BOT
400 A=EXP(YMEAN-XMEAN*B)
410 PRINT 'Slope (alpha) is';B
420 PRINT
430 PRINT ' ****************************************'
440 PRINT ' *                                      *'
450 PRINT ' * Type 1 for Time as a function of MTBF *'
460 PRINT ' * Type 2 for MTBF as a function of Time *'
470 PRINT ' * Type 3 to input another data set     *'
480 PRINT ' * Type 4 to quit                       *'
490 PRINT ' *                                      *'
500 PRINT ' ****************************************'
510 INPUT R$
520 IF R$='1'OR R$='2'OR R$='3'OR R$='4' THEN GOTO 550
530 PRINT ' Input error! Try again'
540 GOTO 420
550 IF R$='1' THEN GOSUB 600
560 IF R$='2' THEN GOSUB 700
570 IF R$='3' THEN GOTO 130
580 IF R$='4' THEN GOTO 800
590 GOTO 420
600 REM
610 REM This GOSUB calculates the time at which a particular
620 REM reliability will be reached
630 PRINT ' Input a MTBF';
640 INPUT MTBF
650 TM=A*MTBF**B/(1-B)
660 PRINT
670 PRINT ' The time is';TM
680 PRINT
690 RETURN
```

94 Estimating reliability

```
700 REM
710 REM this GOSUB calculates the MTBF that will be reached
720 REM after a certain development time
730 PRINT ' Input the time';
740 INPUT TM
750 MTBF=(TM*(1-B)/A)**(1/B)
760 PRINT
770 PRINT ' The MTBF will be';MTBF
780 PRINT
790 RETURN
800 END
```

Program 6.1 Reliability growth analysis using the Duane model

For further reading on this subject see the reference list in Chapter 2. O'Connor (Reference 12) has already been mentioned. Apart from that, Carter (Reference 4) discusses it, and the US publications MIL STD 1635 (Reference 22) and MIL-HDBK 189 (Reference 19) are devoted to it. Def. Stan. 00-41 (Reference 13) also gives up some space to it.

6.2 Uncertainty in results

It is during the development phase of a project that data first appears from the testing of prototypes. It is this data that is used to estimate the reliability of the equipment, the amount of further development needed to attain the specification, and the maintenance requirements of the final system if required by the customer. It is necessary to give a brief description of the statistical tools used to analyse this data.

The data from experiments have a degree of statistical uncertainty, which must then insinuate itself into any results or conclusions that are drawn from it. There are two distinct techniques that will be discussed for incorporating this uncertainty into the results. The first is the classical ideas of significance and confidence limits, and the second approach is that of Bayesian analysis. It is not the purpose of this text to discuss the relative merits of these two approaches. Sufficient to say they are distinct, they both have their disciples, and they both are logically consistent from a mathematical point of view. There are advantages and disadvantages to them both.

6.3 Classical statistical analysis

Significance

Consider the following situation. A manufacturer produces a component of which he claims that no more than 5% are defective.

Probability distributions   95

On testing five of them, one is found to be defective. Is this cause for concern?

The approach to the problem is to ask 'Just how rare is one defective in five, when there is only a 5% probability of each item being defective?' If it is too rare, then one must believe that either a very rare event has taken place, or that the manufacturer is mistaken in his figure of 5%. In actually doing the sum, the 'rareness' is measured by the probability of one defective, or worse, i.e. of one or more defectives. Then, if the manufacturer's claim is true,

$$P \text{ (0 defectives)} = (0.95)^5 = 0.77$$

$$P \text{ (1 or more)} = 1 - 0.77$$

$$= 0.23$$

An event has taken place, the prior probability of which is 23%. This is not terribly infrequent — one could take the manufacturer's word for it.

Notice that this figure is arrived at on the assumption that the 5% is correct. This assumption, which we are testing, is called the null hypothesis, H_0.

In practice, 5% and 1% are often used as 'limiting values' of the probability, called the significance level. If an event takes place that is not expected to occur as often as 5% of the time it is said to be significant at the 5% level, and similarly for the 1% level of significance.

6.4 Probability distributions

Before we can go further, it is necessary to introduce the concept of a probability distribution. This is the language used by statisticians to describe the possible variability in data. The reader will have already met distributions in Chapter 2, although they were not emphasized there.

Data can fall into one of two possible categories — continuous or discrete. Data is discrete if something is being counted (the number of defective items in a sample, or the number of failures of a piece of equipment in a given time period), while it is continuous if something is being measured (time or distance to or between failures). The treatment of distributions is slightly different in the two cases.

Probability distributions of a discrete variable

Suppose the variable, which we call r, can take on the values

$$r = 0, 1, 2, \ldots$$

96   Estimating reliability

Then let $p_0, p_1, p_2, \ldots$ be the probabilities that r takes the values $0, 1, 2, \ldots$ respectively. We can make up a table of the values of r and the corresponding values of the probabilities:

r	0	1	2	3	$\ldots$
p_r	p_0	p_1	p_2	p_3	$\cdots$

This is the probability distribution (the table of all possible values of r and their probabilities). Note that

$$\sum p_r = 1$$

where the sum is taken over all possible values of r.

In practice we shall assume there is a formula for p_r in terms of r, and possibly other parameters as well. There are two such distributions that must be discussed.

The binomial distribution

Suppose I have a sample of n items, chosen at random from a large batch, and it is believed that for each of them there is a probability p of them being defective, and a probability $q = 1 - p$ of their being non-defective. Then it can be shown that if p_r is the probability of exactly r defectives, then

$$p_r = C_r^n p^r q^{n-r}.$$

where

$$C_r^n = \frac{n!}{r!(n-r)!}$$

where

$$n! = n(n-1)(n-2)\ldots 3 \times 2 \times 1$$

Notice that exactly n items are examined, the probability of each one of them being defective is p, and p_r is the probability of exactly r being defective, and hence $n - r$ being non-defective.

The Poisson distribution

The binomial distribution is used when discussing the number of defectives in a sample. The Poisson distribution is used when discussing, for example, the number of breakdowns in a given time period when there is no sample and hence no sample size n, and when, in principle, the number of possible breakdowns is very large and may be unbounded. In this case, if m is the average number of breakdowns that can be expected, then if p_r is the probability of exactly r breakdowns,

$$p_r = e^{-m}\frac{m^r}{r!}$$

The Poisson distribution describes the patterns of failure described above, or any occurrence that happens in a totally random manner, in the sense that the occurrences are independent, i.e. they do not affect each other. The number of road accidents on a stretch of road in a fixed time is a Poisson distribution, because the occurrence or non-occurrence of an accident at a particular place at a particular time is not going to affect the probability of an accident at that spot (or anywhere else) at any time. Failures of complex equipment in a fixed time period have a Poisson distribution only if the failure rate is constant — which it is if we are considering complex equipment.

A further use of the Poisson distribution is that it is a good approximation to the binomial distribution when p is small and n is large, with $m = np$.

Probability distributions of a continuous variable

These have already been mentioned implicitly in the secion on constant failure rate, but in this section they are discussed more explicitly. If t is a continuous variable, like time to failure, its distribution is given by a function, the probability density function (p.d.f.) with the property that

$$P(t_0 < t < t_1) = \int_{t_0}^{t_1} f(t)\,dt$$

= the shaded area under the graph in Figure 6.3.

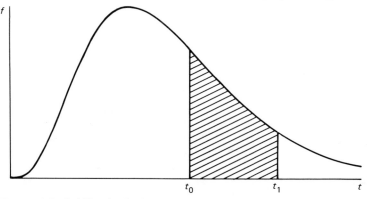

Figure 6.3 Probability density function showing probability as an area

98 Estimating reliability

It has the property that

$$\int f(t)\,dt = 1$$

where the integral is all possible values of t. If t is a time to failure,

$$\int_0^\infty f(t)\,dt = 1$$

because t can, in principle, take any positive value. As with the discrete case, there are two cases in particular that are of interest.

The exponential distribution

$$f(t) = \lambda\,e^{-\lambda t}$$

This has already been discussed in the section on constant failure rate.

The Erlang distribution

$$f(t) = \frac{\lambda^{n+1}}{n!}\,t^n\,e^{-\lambda t}$$

This can be used (though rarely is) to describe times to failure. It was mentioned briefly earlier, and its graph was shown in Figure 6.4. The reason for studying this distribution here is to deal with tests of significance on the MTBF. The procedure is illustrated in the following example.

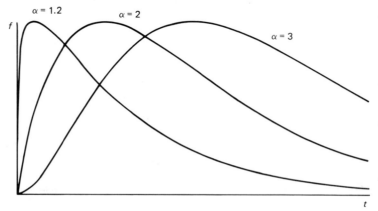

Figure 6.4 The Erland distribution, showing different values of α

Probability distributions   99

A designer believes his design for an item has a MTBF of 100 h under certain specified conditions. Five items are tested until they all fail, and the average of these five times is 85 h. Is this result significant?

Each time five items are tested until failure, their calculated MTBF will be different. This is another way of saying that observed, or experimental, MTBF is a statistical variable, and so has a probability distribution. In this case, with five prototypes tested until failure, the *total running time* (or five times the observed MTBF) is represented by an Erland distribution, with $n = 5$ and λ the failure rate of the components under consideration, i.e. if T is the total time on test, and all the items have failed, the p.d.f. of T is

$$f(T) = \frac{\lambda^6}{6!} T^5 e^{-\lambda T}$$

where $n = 5$ and the null hypothesis is that $\lambda = 1/100 = 0.01$.

The probability of getting a result as poor as $T = 425 = 5 \times 85$ is

$$\int_0^{425} f(T)\, dT$$

This integral can be done by applying the equation for integration by parts four times, or it can be evaluated numerically, as has been done in Program 6.2. The result is

$$\int_0^{425} f(T)\, dT = 0.42$$

which is not at all significant.

Program 6.2, which deals with significance tests, is really three programs in one, as it can deal with Poisson, binomial and exponential distributions data (see the comments in the program itself). For the exponential case, the integral is done using Simpson's rule. In all cases, extreme values of the variables may cause underflow or overflow problems. When transcribing the program, output messages to show the progress as the routine loops can indicate if problems are arising. Using multiprecision variables (if this option is available) distances these problems in some cases.

```
100 REM
110 REM                  SIGNIFICANCE ROUTINE
120 REM
130 REM
140 DATA ' This is significant at the 5% level'
150 DATA ' This is not significant at the 5% level'
160 DATA ' This is significant at the 1% level'
170 DATA ' This is not significant at the 1% level'
180 READ S5$,SN5$,S1$,SN1$
```

100 Estimating reliability

```
190 PRINT ' This routine performs significance tests'
200 PRINT ' The data can be Binomial, Poisson, or Exponential.'
210 INPUT ' Input your choice - B, P, or E',C$
220 PRINT
230 PRINT
240 IF C$ = 'B' THEN 330
250 IF C$ = 'P' THEN 650
260 IF C$ = 'E' THEN 930
270 GOTO 210
280 INPUT ' Do you want to analyse more data (Y/N)',A$
290 IF A$ = 'Y' THEN 210
300 IF A$ = 'N' THEN STOP
310 PRINT ' Response is Y (yes) or N (no) '
320 GOTO 280
330 REM
340 REM                    BINOMIAL DATA
350 REM
360 INPUT ' Sample size',N
370 INPUT ' Number of failures',K
380 INPUT ' Hypothesised value of the proportion of defectives',P
390 PRINT
400 PRINT
410 S=0
420 IF K = 0 THEN 550
430 Q=1-P
440 COEF=1
450 FOR I=0 TO K-1
460 PWR=1
470 FOR J=1 TO N
480 IF J > I THEN PWR=PWR*Q ELSE PWR=PWR*P
490 NEXT J
500 S=S+COEF*PWR
510 COEF=COEF*(N-I)/(I+1)
520 NEXT I
530 S=1-S
540 PRINT ' The value of P is ',S
550 IF S <= .05 THEN PRINT S5$ ELSE PRINT SN5$
560 IF S <= .01 THEN PRINT S1$ ELSE PRINT SN1$
570 PRINT
580 PRINT
590 INPUT ' Do you want to analyse more binomial data (Y/N)',A$
600 IF A$ = 'N' THEN 280
610 PRINT
620 IF A$ = 'Y' THEN 330
630 PRINT ' Response is Y (yes) or N (no) '
640 GOTO 590
650 REM
660 REM                    POISSON DATA
670 REM
680 INPUT ' Number of failures ',K
690 INPUT ' Hypothesised mean ',M
700 PRINT
710 PRINT
720 S=0
730 IF K=0 THEN 820
740 COEFF=EXP(-M)
750 S=COEFF
760 IF K=1 THEN 760
770 FOR I=1 TO K-1
780 COEFF=COEFF*M/I
```

Probability distributions 101

```
790 S=S+COEFF
800 NEXT I
810 S=1-S
820 PRINT ' The value of P is ',S
830 IF S <= .05 THEN PRINT S5$ ELSE PRINT SN5$
840 IF S <= .01 THEN PRINT S1$ ELSE PRINT SN1$
850 PRINT
860 PRINT
870 INPUT ' Do you want to analyse more Poisson data (Y/N)',A$
880 IF A$ = 'N' THEN 280
890 PRINT
900 IF A$ = 'Y' THEN 650
910 PRINT ' Response is Y (yes) or N (no) '
920 GOTO 870
930 REM
940 REM              EXPONENTIAL DATA
950 REM              INTEGRAL BY SIMPSON
960 REM
970 INPUT 'Number of items on test',N
980 INPUT 'Hypothesised MTTF',LAM
990 PRINT
1000 S=0
1010 PRINT 'Now input the failure times'
1020 FOR I = 1 TO N
1030 INPUT ' Failure time ',F
1040 S=S+F
1050 NEXT I
1060 PRINT
1070 PRINT
1080 H=S/10
1090 MX=5
1100 OI=0
1110 F=1./LAM
1120 FOR J = 1 TO N-1
1130 F=F/J/LAM
1140 NEXT J
1150 LAST=F*EXP(-S/LAM)*S**(N-1)
1160 EVENS=0
1170 FOR I = 1 TO MX-1
1180 T=2*I*H
1190 EVENS=EVENS+F*EXP(-T/LAM)*T**(N-1)
1200 NEXT I
1210 ODDS=0
1220 FOR I = 1 TO MX
1230 T=(2*I-1)*H
1240 ODDS=ODDS+F*EXP(-T/LAM)*T**(N-1)
1250 NEXT I
1260 NI=(LAST+4*ODDS+2*EVENS)*H/3
1270 REM TEST FOR CONVERGANCE
1280 IF ABS(NI-OI) < .0001 THEN 1340
1290 EVENS=ODDS+EVENS
1300 OI=NI
1310 H=H/2
1320 MX=MX*2
1330 GOTO 1210
1340 NI=1-NI
1350 PRINT ' The value of P is ',NI
1360 IF NI <= .05 THEN PRINT S5$ ELSE PRINT SN5$
1370 IF NI <= .01 THEN PRINT S1$ ELSE PRINT SN1$
1380 PRINT
```

102 Estimating reliability

```
1390 PRINT
1400 INPUT ' Do you want to analyse more Exponential data',A$
1410 IF A$ = 'N' THEN 280
1420 PRINT
1430 IF A$ = 'Y' THEN 930
1440 PRINT ' Response is Y (yes) or N (no) '
1450 GOTO 1400
```

Program 6.2 Significance tests

This raises the question of rejecting or accepting the hypothesis. In practice, the significance level is chosen in advance, and the accept/reject decision made by comparing p with this level. The value of the significance level chosen must depend on the relative costs of making a mistake. There are two possible mistakes that can be made. These are

- Type I error — rejecting a true hypothesis as false or
- Type II error — accepting a false hypothesis as true.

Clearly, the cost of committing one of these errors, and the likelihood of their occurrence, will depend on the situation, though the probability of committing a Type I error is the level of significance chosen. For a manufacturer of safety equipment, the cost of a Type II error (accepting a false hypothesis) could involve loss of life, and/or expensive compensation, and he may be willing to risk making Type I errors (rejecting a true hypothesis, or rejecting good stuff) in order to avoid that situation. A manufacturer of cheap domestic equipment, on the other hand, such as alarm clocks, may be more willing to accept the risk of producing poor goods, as the cost of replacing them and compensating irate customers is less than that of rejecting good material. This book is not suggesting that manufacturers should, or do, deliberately put faulty goods on the market. Every reasonable effort should be made to improve quality and reliability, but we live in a cost-conscious world, and these tradeoffs have to be made.

6.5 Confidence intervals

Consider the following problem. A sample of ten items are tested and all of them are found to be sound. Let p be the proportion of defective items in the total population of such items. Then what can be said about p, and the possible values it can take?

Let $q = 1 - p$ be the proportion of good items. The estimate of q is 1, but clearly nobody would believe that p is really zero. How large

Confidence intervals   103

can p reasonably be? Or alternatively, how small can q be? Clearly if

$$q^{10} > 0.05$$

i.e.

$$q < (0.05)^{1/10}$$
$$= 0.74$$

or

$$p < 0.26$$

then the result would not be significant, in the sense that if p took on a value greater than 0.26 then the number of defectives, at zero, would be significantly low at the 5% significance level. The value limiting value of p, 0.26, is called the 95% upper confidence level on p. In the examples that follow, we shall normally be interested in the upper confidence limit, because we are interested in the largest reasonable value that p may take. We shall normally omit the adjective 'upper'.

The principle behind confidence limits is that we are seeking the limiting value of a parameter, beyond which the observed result would be significant at a prescribed level α. In the three cases considered, we are interested in keeping the value of the parameter low (proportion of defective items in a population, the average number of failures in a fixed time period, failure rate), and so we have considered the upper confidence limit. In the two cases when we are interested in keeping the value high (MTTF, reliability), we consider lower confidence limits, but in the latter case, the confidence limit is a simple function of the limit in one of the former cases.

The technique, illustrated in the examples above, is to consider the probability of a result as extreme, or more extreme, than the one observed. Call this probability P, and consider it as a function of the parameter of interest, x. Then solve

$$P < \alpha$$

for x. In our cases, if x_u is the solution of

$$P = \alpha$$

then

$$P < \alpha$$

if

$$x > x_u$$

and x_u is called the $(1 - \alpha)$ confidence limit of x. x_u is the largest value of x for which the observed result is not significant at the α level.

104   Estimating reliability

Program 6.3 is very similar to the previous one, except that in this case the integration is done analytically.

```
100 REM
110 REM                CONFIDENCE LIMIT ROUTINE
120 REM
130 PRINT ' This routine calculates confidence limits '
140 PRINT ' The data can be Binomial, Poisson or Exponential '
150 INPUT ' Input your choice B, P or E ',A$
160 PRINT
170 PRINT
180 IF A$ = 'B' THEN 270
190 IF A$ = 'P' THEN 600
200 IF A$ = 'E' THEN 770
210 GOTO 150
220 INPUT ' Do you want to analyse more data ',A$
230 IF A$ = 'Y' THEN 150
240 IF A$ = 'N' THEN STOP
250 PRINT ' The response is Y(yes) or N(no)'
260 GOTO 150
270 REM
280 REM                BINOMIAL DATA
290 REM
300 INPUT ' Sample size ',N
310 INPUT ' Number of failures ',R
320 Input ' Confidence level ',AL
330 PRINT
340 PRINT
350 P=(2*R+1-SQRT((2*R+1)*(2*R+1)-4*(N+1)*R*R/N))/(1+N)/2
360 Q=1-P
370 DEN=Q**N
380 TERM=N
390 FOR I=1 TO R
400 DEN=DEN+TERM*Q**(N-2*I+1)*P
410 TERM=TERM*Q*P*(N-I)/(I+1)
420 NEXT I
430 DEN=DEN-1+AL
440 NM=Q**(N-R-1)
450 FOR I=1 TO R
460 NM=NM*P*(N-I)/I
470 NEXT I
480 PN=P+DEN/NM/N
490 IF ABS(PN-P) < .0001 THEN 520
500 P=PN
510 GOTO 360
520 PRINT ' The upper limit on the proportion of defectives is ',PN
530 PRINT
540 PRINT
550 INPUT ' Do you want to analyse more binomial data (Y/N)',A$
560 IF A$ = 'N' THEN 220
570 IF A$ = 'Y' THEN 270
580 PRINT ' The response is Y(yes) or N(no)'
590 GOTO 550
600 REM
610 REM                POISSON DATA
620 REM
630 INPUT ' Number of incidents ',N
640 INPUT ' Confidence level ',AL
650 PRINT
```

Confidence intervals   105

```
660 PRINT
670 MN=-LOG(1-AL)
680 IF N > 0 THEN GOSUB 1030
690 PRINT ' The upper limit on the mean number of incidents is ',MN
700 PRINT
710 PRINT
720 INPUT ' Do you want to analyse more Poisson data',A$
730 IF A$ = 'N' THEN 220
740 IF A$ = 'Y' THEN 600
750 PRINT ' The response is Y(yes) or N(no)'
760 GOTO 720
770 REM
780 REM             EXPONENTIAL DATA
790 REM
800 INPUT ' Number of readings ',N
810 INPUT ' Confidence level ',AL
820 PRINT ' Input the failure times '
830 T=0
840 FOR I=1 TO N
850 INPUT ' failure time ',X
860 T=T+X
870 NEXT I
880 PRINT
890 PRINT
900 MN=-LOG(1-AL)
910 N=N-1
920 IF N > 0 THEN GOSUB 1030
930 LAM=MN/T
940 PRINT ' The upper limit on the failure rate is ',LAM
950 PRINT ' The lower limit on the MTTF is ',1/LAM
960 PRINT
970 PRINT
980 INPUT ' Do you want to analyse more exponential data ',A$
990 IF A$ = 'N' THEN 220
1000 IF A$ = 'Y' THEN 770
1010 PRINT ' The response is Y(yes) or N(no) '
1020 GOTO 970
1030 REM
1040 REM                    ITTERATIVE PROCEDURE
1050 REM
1060 M=N
1070 MN=0
1080 TERM=1
1090 SUM=TERM
1100 FOR I=1 TO N
1110 TERM=TERM*M/I
1120 SUM=SUM+TERM
1130 NEXT I
1140 M=LOG(SUM/(1-AL))
1150 IF ABS(MN-M) < .0001 THEN RETURN
1160 MN=M
1170 GOTO 1080
```

Program 6.3 Confidence limits

Most elementary texts on statistics deal with the ideas of
significance and confidence. For a deeper examination of the
concepts involved see the references given in Chapter 2. Ott
(Reference 13) and DeGroot (References 5 and 6), give a good

106   Estimating reliability

exposition. For some further applications in the reliability field, Mann (Reference 10), Gnedenko (Reference 7), Lloyd (Reference 9) and Bain (Reference 1) are to be recommended.

6.6 Bayesian estimation

An alternative approach to the idea of confidence limits and significance tests is that of Bayesian analysis. Before we can begin to discuss estimation, we must discuss Bayes' theorem — named after the first person to publish it, an eighteenth century vicar, Thomas Bayes.

Bayes' theorem

It is easiest to demonstrate the theorem by the use of examples. Suppose that a builder buys bolts from a retailer. He knows the retailer buys his bolts from two manufacturers, A and B, and he knows from long experience of testing these bolts that on average 1/200 bolts from A fail the test while 1/250 from B fail. He sends the office boy out for some bolts from the retailer, and on his return the office boy reveals that he did not record the type of bolts purchased. The builder tests 1000 of the bolts and six fail. What can we say about who manufactured them? In particular, what is the probability that they are A's bolts as opposed to B's?

The average number of bolts to fail in a sample will be five if they are from A and eight if they are from B. Using this and the Poisson distribution

$$P \text{ (6 fail if A's bolts)} = 0.15$$

$$P \text{ (6 fail if B's bolts)} = 0.12$$

It is reasonable to assume, and indeed can be proved mathematically, that P (the bolts are from A) and P (the bolts are from B) are proportional to these values. This gives

$$P \text{ (they are from A)} = 0.15/(0.15 + 0.12)$$

$$= 0.55$$

$$P \text{ (they are from B)} = 0.12/(0.15 + 0.12)$$

$$= 0.45$$

because the probabilities must add up to 1. Therefore, it is slightly more likely that the bolts came from A.

The office boy then shows that he is not as stupid as he looks, and points out that the retailer in fact stocks A's bolts four times more often than those of B, i.e. the probability that the bolts were from A is 0.8 anyway. Will this make a difference to our assessment? Can we use this information as well as that of the test?

The initial information, that

$$P \text{ (they are from A)} = 0.8$$

and so

$$P \text{ (they are from B)} = 0.2$$

is called prior information, and the values denoted by π. So write

$$\pi(A) = 0.8$$

$$\pi(B) = 0.2$$

Then Bayes' theorem says that

$$P \text{ (they are from A)} \propto 0.15 \times 0.8 = 0.120$$

$$P \text{ (they are from B)} \propto 0.12 \times 0.2 = 0.024$$

and again using the fact that the probabilities must add up to 1, this gives

$$P \text{ (they are from A)} = 0.83$$

$$P \text{ (they are from B)} = 0.17$$

This is an example of Bayes' theorem, which in mathematical symbols is written (in this case)

$$P \text{ (A|6 fail)} \propto P \text{ (6 fail|A)}\pi(A)$$

where

$P \text{ (A|6 fail)}$ is the probability that the bolts are from manufacturer A given that we have observed the six failures in the 1000 that were tested.

$P \text{ (6 fail|A)}$ is the probability of six of A's bolts failing when 1000 are tested.

These probabilities, involving two events, of which we have knowledge of one, are called conditional probabilities, as the probability of one event is conditional on our knowledge of the other.

$\pi(A)$ is the probability that the bolts come from A, irrespective of the results of the test (or before we have tested). This is called the marginal probability, or in the situation we are going to study, the prior probability.

Notice that the theorem is given in terms of proportionality. Because the probabilities must sum to one we can then calculate the actual probabilities.

108　Estimating reliability

Estimation

In the example above and Worked Example 6.11 there were two variables that were considered, along with the possible ways they could effect each other. In the first example the variables were the number of failed bolts observed and the identity of the manufacturer, and in the second example they were whether or not a failure was present, and whether or not a failure was (apparently) detected.

To use Bayes' Theorem for estimation, the two variables are the data we observe and the parameter that is being estimated. Uncertainty about the parameter is expressed using statistical ideas. We can consider the probability that a parameter is greater than a given value, for example. In particular, we can express our uncertainty about the value of a given parameter by putting a distribution on it. Bayes' theorem in that case is

$$P \text{ (parameter|data)} \propto P \text{ (data|parameter)} \, \pi \text{ (parameter)}$$

The terms of this expression are:

1. *P (parameter|data)*.　This is the final probability distribution on the parameter, and obviously is a function of the data. In giving our results in this way, we can put probability intervals on the parameter instead of confidence limits. This distribution is called the posterior distribution.
2. *P (data|parameter)*.　This is the probability of getting the data in question as a function of the parameter, and depends on the probability distribution of, for example, the time to failure of a component. It is called the likelihood function.
3. *P (parameter)*.　This is the prior, or marginal, distribution on the parameter, and is denoted π. It comes from previous experience and data, and represents the experimenter's knowledge and uncertainty about the parameter. It may well be the posterior from previous experiments.

If a point estimate of the parameter is required, very often the value that maximizes the posterior function is used. This is called the maximum likelihood estimate, or MLE.

The reader will observe that the major difference between this case and the two examples discussed above is the use of probability distributions, rather than the probabilities of discrete events. This involves a minimum of further complications, simply that when normalizing to calculate the posterior, is it necessary to integrate overall the possible values of the parameter concerned, rather than just taking the sum over them, as was done previously. In order to be able to discuss the analysis, it becomes necessary to discuss two more distributions. The reader may like to omit this section initially, and return to it later, or refer to it as necessary when studying the applications.

6.7 Further distributions

The gamma distribution

Put

$$\Gamma(\alpha) = \int_0^\infty x^{\alpha-1}\, e^{-x}\, dx$$

$\Gamma(\alpha)$ (pronounced gamma alpha) is called the gamma function. It is an extension of the factorial function to non-integral values, in the sense that it is a well-defined continuous function for all values of α except a non-positive integer, and

$$\Gamma(a) = (\alpha - 1)\Gamma(\alpha - 1)$$

and

$$\Gamma(\alpha) = (\alpha - 1)!$$

when α is a positive integer. In general, $\Gamma(\alpha)$ can only be evaluated by using numerical techniques on a computer (a polynomial approximation is used in the program at the end of the section, valid between 1 and 2, and the relationship above is used to extend the function beyond this range). Some values can be calculated analytically, and it can be shown, for example that

$$\Gamma(1/2) = \sqrt{\pi}$$

The gamma function is used in the following manner. If x is a random variable with probability density function

$$f(x) = \frac{\theta^\alpha}{\Gamma(\alpha)}\, x^{\alpha-1}\, e^{-\theta x}$$

then x is said to be gamma distributed with shape parameter α and scale parameter θ. Note that this is very similar to the Erlang distribution, but the shape parameter can now take on non-integral values. The term $\theta^\alpha/\Gamma(\alpha)$ is a normalizing term that ensures the integral is one. (The reader may like to verify this. It is a simple exercise in integration. The definition of the gamma function must be assumed.) It is not difficult to show that the mode of x (where $f(x)$ has its maximum) is at

$$x = \frac{\alpha - 1}{\theta}$$

In certain cases, the posterior distribution of the parameter of interest will be a gamma distribution, in which case the mode will be the MLE.

The beta distribution

Put

$$B(\alpha, \beta) = \int_0^1 x^{\alpha-1}(1-x)^{\beta-1}\, dx$$

$B(\alpha, \beta)$ (pronounced beta alpha, beta) is called the beta function. Like the gamma function, it is extensively used in statistical work. There will be little further discussion here, except to observe that

$$B(\alpha, \beta) = \frac{\Gamma(\alpha)\Gamma(\beta)}{\Gamma(\alpha+\beta)}$$

which is the formula used in the final program to evaluate $B(\alpha, \beta)$.

If x is a random variable, with p.d.f.

$$f(x) = \frac{x^{\alpha-1}(1-x)^{\beta-1}}{B(\alpha, \beta)} \qquad 0 \leqslant x \leqslant 1$$

then we say that x is beta distributed with parameters α and β. Two examples of this function are shown in Figure 6.5. The mode is at

$$x = \frac{\alpha - 1}{\alpha + \beta - 2}$$

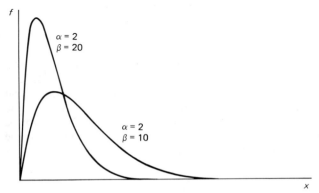

Figure 6.5 The beta distribution, showing different values of α and β

Further distributions   111

Estimating a constant failure rate

In this instance, the data will be times to failure, and we wish to estimate λ, the MTTF, or equivalently, the failure rate λ. The p.d.f. of the time to failure is

$$f(t) = \lambda\,e^{-\lambda t}$$

and this will be used to calculate the likelihood function. The technique is best illustrated by the use of examples.

Four similar items are tested until failure, and before the experiment the engineer is questioned to determine his appraisal of the failure rate (assumed constant) of the item. His estimate of the failure rate is summarized in terms of a probability distribution on the failure rate λ. This is the prior information, and he believes it to be

$$\pi(\lambda) = \frac{100^4}{2!}\,\lambda^2\,e^{-100\lambda}$$

(The means by which this may be done is discussed below.) The data is that all of the items failed, at times 23, 25, 34 and 41 hours.

If λ is the failure rate of the item under test, then the probability of the first failing at precisely 25 h is proportional to the p.d.f. evaluated at $t = 23$, i.e. to

$$\lambda\,e^{-23\lambda}$$

and similarly for the other three items. Then the probability of all these failure times being observed is proportional to the product of these, namely

$$\lambda^4\,e^{-23\lambda}\,e^{-25\lambda}\,e^{-34\lambda}\,e^{-41\lambda} = \lambda^4\,e^{-123\lambda}$$

which is the likelihood function,

$$L(\lambda) = \lambda^4\,e^{-123\lambda}$$

This gives the posterior distribution

$$P(\lambda) \propto \lambda^2\,e^{-100\lambda} \times \lambda^4\,e^{-123\lambda}$$
$$= \lambda^6\,e^{-223\lambda}$$

The MLE of λ is

$$\lambda = 0.030$$

and that of the MTTF is the reciprocal of this,

$$\text{MTTF} = 1/0.030$$
$$= 37.2$$

The 95% upper Bayesian limit of λ, i.e. the value λ_μ, such that

$$P(\lambda < \lambda_\mu) = 0.95$$

is the value of λ such that

$$\int_0^{\lambda_\mu} P(\lambda)\,d\lambda = 0.95$$

The Program 6.4 at the end of this section solves this equation, the solution of which is

$$\lambda_\mu = 0.053$$

These values are shown on the graph in Figure 6.6.

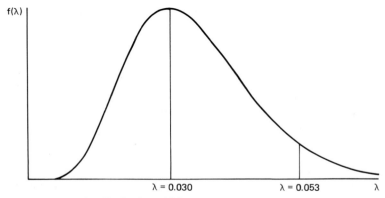

Figure 6.6 Posterior distribution of failure rate

The lower 95% Bayesian limit on the MTTF is the reciprocal of this,

$$= 1/0.053$$
$$= 18.8$$

The reader will observe that in this case the prior information on λ was a gamma distribution, which led to a gamma distribution as posterior. There is no good engineering reason why $\pi(\lambda)$ should be gamma (as opposed to, for example, log normal or Weibull), but the analytic advantage is one that cannot be ignored. In this case the gamma distribution is said to be conjugate, or a conjugate prior. In practice (though not always) the engineer's ideas on the prior information are usually sufficiently vague to allow them to be fitted (or approximated) by a conjugate prior. This is illustrated in the example below. The remainder of this section and Program 6.4 only discuss the use of the appropriate conjugate prior.

Five items are tested, but the trial cannot go on until they all fail. As a result, three of them fail at times 61, 33 and 73 hours. The experiment had to be stopped when the two remaining items had been tested without failure for 90 hours. These are called censored data. The engineers agreed prior to the experiment that the most likely value of the failure rate was 0.01, and they were 95% certain that λ was less than 0.025. Assuming a conjugate prior, i.e. one of the form

$$\pi(\lambda) \propto \lambda^{\alpha-1} e^{-\theta\lambda}$$

for some value of α and θ, it can be shown that

$$\alpha = 4.2$$

and

$$\theta = 323.5$$

This is shown in Figure 6.7.

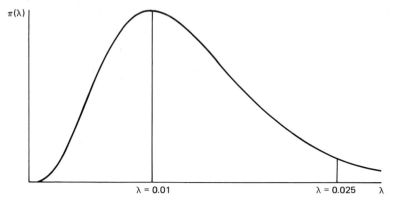

Figure 6.7 Prior distribution of failure rate

The likelihood function in this case is made slightly more complicated by the fact that there are two censored times. As the likelihood function is proportional to the probability of getting the data that was actually observed, it must be

$$L(t|\lambda) \propto f(61)f(68)f(78)R(90)R(90)$$
$$= \lambda^3 e^{-387\lambda}$$

as the probability of a single component still functioning after 90 h is

$$R(90) = e^{-90\lambda}$$

This gives the posterior

$$P(\lambda) \propto L(t)\pi(\lambda)$$

$$P(\lambda) = \frac{(710.5)^{8.2}}{\Gamma(7.2)} \lambda^{7.2} e^{-710.5\lambda}$$

i.e. $P(\lambda)$ is a gamma function with shape parameter 7.2 and scale parameter 710.5. The program can now calculate the MLE and λ_μ, as well as those of the MTTF, and these are given by

$$\text{MLE}(\lambda) = 0.009$$

$$\lambda_\mu = 0.017$$

and

$$\text{MLE(MTTF)} = 114.0$$

$$\text{MTTF}_1 = 58.5$$

Notice that the experiment's idea of the most likely value of λ and the MTTF has not been changed. However the 95% Bayesian limit has changed. This can be interpreted as saying he is now more certain of the result. (It would be appropriate to use the word confident here, rather than certain, but this is already a technical statistical term.) The results for λ are shown in Figure 6.8, which shows the posterior distribution. The reader should compare this with Figure 6.7, which shows the prior.

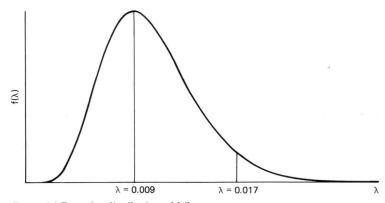

Figure 6.8 Posterior distribution of failure rate

Further distributions 115

Estimating the mean of a Poisson distribution

In this case a number of events (breakdowns in a fixed time period, for example) are observed. If this number is believed to be part of a Poisson distribution, then

$$P \text{ (number} = r) = e^{-m}\frac{m^r}{r!}$$

and it is necessary to estimate m, the mean. The conjugate prior is again gamma.

Estimating a proportion

In this section we shall consider the problem of estimating the proportion of defective items in a population. The conjugate prior on p, the proportion of defectives, is beta in this case. An example is given below.

A manufacturer believes that a particular manufacturing process results in about 2% of the components produced being defective, though he believes it could be as high as 10% (the upper 95% limit). Twenty items are examined, and one of them is found to be defective. How does this affect his estimate?

The prior is beta, so

$$\pi(p) = \frac{p^{\alpha-1}(1-p)^{\beta-1}}{B(\alpha, \beta)}$$

for some α and β. As the mode of p is 0.02, and the 95% upper limit is 0.1, then

$$\int_0^{0.1} \pi(p)\,dp = 0.95$$

as shown in Figure 6.9

This gives values of 1.8 and 42.1 as values of α and β respectively. The likelihood function is

$$L = p^1(1-p)^{19}$$

because the number of defectives observed is binomial. This gives

$$P(p) = \frac{p^{1.8}(1-p)^{60.1}}{B(2.8, 61.1)}$$

116 Estimating reliability

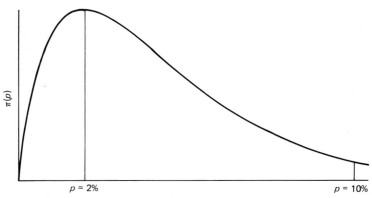
Figure 6.9 Prior distribution of a proportion

as the posterior, i.e. the posterior is beta with 2.0 and 61.1 as α and β respectively. This gives

$$\text{MLE}(p) = 0.030$$

and the 95% upper Bayesian limit

$$p_\mu = 0.093$$

The posterior distribution is shown in Figure 6.10. The reader should compare this with Figure 6.9, which shows the prior information.

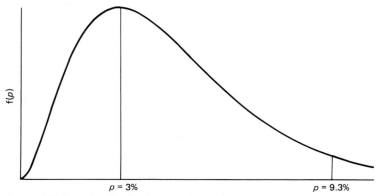
Figure 6.10 Posterior distribution on a proportion

Bayesian analysis

Bayesian analysis is a very powerful technique, because the experimenter is essentially crediting himself with data, summarized in the prior information, before the start of the experiment. For example, in the last example above, the analysis was equivalent to having already observed 1.8 defectives and 42.1 non-defectives (the analysis does not object to the fractions in this case). The prior information can come from previous data if required, which may be 'doctored' to allow for changes since it was collected. The following example illustrates this.

Consider an example of the lorries and the number of breakdowns in a week (see Worked Example **(6.12)**). The manager of the fleet allows more money to be spent on preventative maintenance, which it is believed will reduce the average number of breakdowns from 1.7 to 1.2. They are, however, no more certain of the value, so the 95% limit remains at 2.5. This gives a prior distribution of

$$\pi(m) = \frac{(4.3)^{7.1}}{\Gamma(6.1)} \, m^{6.1} \, e^{-4.5m}$$

i.e. gamma with shape parameter 6.1 and scale parameter 4.3. During the following week, a single breakdown is observed. This gives

$$L = m \, e^{-m}$$

and the posterior

$$P = \frac{(5.3)^{8.1}}{\Gamma(7.1)} \, m^{7.1} \, e^{-5.3m}$$

i.e. gamma with shape parameter 7.1 and scale parameter 5.3. The MLE and 95% upper limit are then

$$MLE(m) = 1.4$$

$$m_\mu = 2.3$$

The manager of the production plant considered above is not happy with the quality of the goods produced. (MLE of 3% and 95% limit of 9.3%.) Changes are made to the production process that it is believed will bring p down to 1%, but in any case to less than 3% (the upper 95% limit). How large a sample should be used to verify this, if the MLE is to be 1% but the upper limit reduced to 2%, and no defectives are observed?

The prior information is beta with values 2.8 and 61.1 for α and β respectively. The program will not solve this problem directly, but

118   Estimating reliability

the sample size must be found by trial and error. The following table sums up the results.

Trial sample	Alpha	Beta	MLE	95% Upper limit
50	2.8	111.1	0.016	0.052
100	2.8	161.1	0.011	0.036
150	2.8	211.1	0.008	0.027
200	2.8	261.1	0.007	0.023

It is relatively easy to obtain the MLE of 1%, but the assurance that the upper limit is less than 2% is somewhat more expensive!

Program 6.4 in this section is the longest and probably the most complicated in the book. Again it is in three sections, for Poisson, binomial and exponential data. Initially the program has to solve the integral

$$\int_0^x f(x)\,dx = 0.95$$

in order to find the parameters of the function f (the prior distribution) for a fixed value of x. When it has the data, and can calculate the parameters of the posterior, and then has to solve the same integral for x for fixed parameters. Linear interpolation is used to solve the equations, the integrals being calculated by Simpson's rule. The values of the gamma function are calculated using a polynomial approximation, the coefficients of which are in DATA statements at the start of the program. Extreme values of the variables may crash the program, or possibly make it loop indefinitely. Using multiprecision variables, if they can be implemented, can help to solve this problem.

Program 6.4 is constructed to be as user friendly as possible, and it requires prompts at all stages. The reader is advised to persist in getting the program running because it will lead to a deeper understanding of the Bayesian analysis of reliability data.

```
100 REM
110 REM
120 DATA .035868343,-.193527818,.482199394,-.756704078
130 DATA .918206857,-.897056937,.988205891,-.577191652
140 DATA '*************************************************'
150 REM
160 DIM GAMMA(8)
170 FOR I=1 TO 8
180 READ GAMMA(I)
190 NEXT I
200 READ RES$
210 PRINT
220 PRINT RES$
230 PRINT '*                                          *'
```

Further distributions 119

```
240 PRINT '*    This routine performs simple Bayesian data analysis  *'
250 PRINT '*    on binomial, Poisson or exponential data.             *'
260 PRINT '*                                                          *'
270 PRINT RES$
280 PRINT
290 PRINT
300 A3=0
310 TH3=0
320 PRINT 'What is your data?'
330 INPUT 'Binomial(B),Poisson(P) or exponential(E)',R$
340 IF R$ = 'B' THEN GOTO 2080
350 IF R$ = 'E' OR R$ = 'P' THEN GOTO 390
360 PRINT 'Input B (binomial), P (Poisson) or E (exponential).'
370 PRINT 'Try again.'
380 GOTO 300
390 REM This section deals with expontial and Poisson data.
400 PRINT
410 INPUT '                     Do you have a prior',RP$
420 IF RP$ = 'Y' THEN GOTO 460
430 IF RP$ = 'N' THEN GOTO 1190
440 PRINT 'The response is Y(yes) or N(no). Try again.'
450 GOTO 410
460 PRINT
470 PRINT 'The program assumes the prior is a gamma distribution'
480 PRINT 'with shape parameter alpha and scale parameter theta.'
490 PRINT 'If you have your own values of alpha and theta they can'
500 PRINT 'be input, otherwise values can be calculated from prior'
510 PRINT 'information.'
520 PRINT
530 INPUT ' Do you have values for alpha and theta',RV$
540 IF RV$ = 'Y' THEN GOTO 1150
550 IF RV$ = 'N' THEN GOTO 600
560 PRINT 'The response is Y(yes) or N(no). Try again.'
570 GOTO 530
580 REM This section calculates the values of alpha and theta from
590 REM prior information input by the operator
600 PRINT
610 PRINT 'It is now assumed that you know the most likely value of'
620 IF R$ = 'P' THEN GOTO 670
630 PRINT 'lambda, the failure rate, and the upper 95% limit (i.e. the'
640 PRINT 'value lmax such that there is a 95% probability that lambda'
650 PRINT 'is less than lmax).'
660 GOTO 690
670 PRINT 'm, the mean, and the upper 95% limit (i.e. the value m-max'
680 PRINT 'such that there is a 95% probability that m is less than m-max).'
690 PRINT
700 INPUT ' What is the most likely value ',LAMM
710 INPUT ' and the 95 % upper limit ',LAMU
720 AL =.05
730 A2=2
740 TERM=EXP(-LAMU*(A2-1)/LAMM)
750 SUM=TERM
760 FOR I=1 TO A2-1
770 TERM=TERM*(A2-1)*LAMU/LAMM/I
780 SUM=SUM+TERM
790 NEXT I
800 IF SUM < AL THEN 840
810 SUM1=SUM
820 A2=A2+1
830 GOTO 740
840 TH2=(A2-1)/LAMM
850 SUM2=1-SUM
860 SUM1=1-SUM1
870 A1=A2-1
880 TH1=(A1-1)/LAMM
890 REM  Start of the loop to find the prior values of alpha and theta
```

120 Estimating reliability

```
900  D=(.95-SUM1)/(SUM2-SUM1)
910  A3=A1+(A2-A1)*D
920  TH3=TH1+(TH2-TH1)*D
930  X=A3-1
940  TPA=TH3**A3
950  GOSUB 4320
960  GOSUB 3850
970  IF ABS(SUM3-.95) < .0001 THEN GOTO 1070
980  IF SUM3 < .95 THEN GOTO 1030
990  A2=A3
1000 TH2=TH3
1010 SUM2=SUM3
1020 GOTO 890
1030 A1=A3
1040 TH1=TH3
1050 SUM1=SUM3
1060 GOTO 890
1070 PRINT
1080 PRINT RES$
1090 PRINT
1100 PRINT '         Prior values, Alpha equals ',A3
1110 PRINT '                      Theta equals ',TH3
1120 PRINT
1130 PRINT RES$
1140 GOTO 1190
1150 REM This section input known values of alpha and theta
1160 PRINT
1170 INPUT 'What is your value of alpha',A3
1180 INPUT '                   and theta',TH3
1190 REM
1200 REM This section inputs exponential data for analysis
1210 REM
1220 PRINT
1230 PRINT 'You must now input the data you wish to analyse. The'
1240 IF R$ = 'P' THEN GOTO 1500
1250 PRINT 'inputs consist of the total number of items on test,'
1260 PRINT 'the number that have failed, the times to failure,'
1270 PRINT 'and the censored times.'
1280 PRINT
1290 INPUT ' How many items alltogether on test ',NT
1300 INPUT '          and how many of them failed ',NF
1310 IF NF <= NT THEN GOTO 1340
1320 PRINT 'There must be no more failures than were tested'
1330 GOTO 1290
1340 PRINT
1350 PRINT 'Input the times. First the failure times, one at a time.'
1360 TFT=0
1370 PRINT
1380 FOR I=1 TO NF
1390 INPUT 'Failure time',TF
1400 TFT=TFT+TF
1410 NEXT I
1420 PRINT
1430 PRINT 'and the censored times.'
1440 PRINT
1450 FOR I=1 TO NT-NF
1460 INPUT 'Censored time',TF
1470 TFT=TFT+TF
1480 NEXT I
1490 GOTO 1550
1500 PRINT 'input is the number of failures observed in the fixed time period.'
1510 PRINT
1520 INPUT 'How many failures ',NF
1530 TFT=1
1540 REM
1550 REM Now the data is input, the MLE and upper 95% limit
```

Further distributions 121

```
1560 REM can be calculated
1570 A3=A3+NF
1580 TH3=TH3+TFT
1590 PRINT
1600 PRINT RES$
1610 PRINT
1620 PRINT '    Posterior values, Alpha equals ',A3
1630 PRINT '                        Theta equals ',TH3
1640 PRINT
1650 PRINT RES$
1660 LAMM=A3/TH3
1670 PRINT
1680 PRINT RES$
1690 PRINT
1700 PRINT '          The most likely value is',LAMM
1710 IF R$ = 'E' THEN PRINT '              and of the MTTF is',1/LAMM
1720 NL=1
1730 LAMU=LAMM
1740 X=A3-1
1750 TPA=TH3**A3
1760 GOSUB 4320
1770 GOSUB 3850
1780 IF SUM3 > .95 THEN GOTO 1840
1790 SUM1=SUM3
1800 NL=NL+1
1810 LAMU=LAMU+LAMM
1820 GOTO 1770
1830 REM LAMM between (NL-1)*LAMM and (NL*LAMM
1840 LAM1=(NL-1)*LAMM
1850 SUM2=SUM3
1860 LAM2=NL*LAMM
1870 LAMU=LAM1+(LAM2-LAM1)*(.95-SUM1)/(SUM2-SUM1)
1880 GOSUB 3850
1890 IF ABS(SUM3-.95) < .0001 THEN GOTO 1970
1900 IF SUM3 < .95 THEN GOTO 1940
1910 SUM2=SUM3
1920 LAM2=LAMU
1930 GOTO 1870
1940 SUM1=SUM3
1950 LAM1=LAMU
1960 GOTO 1870
1970 PRINT
1980 PRINT '   The 95% upper Bayesian limit is',LAMU
1990 IF R$ = 'E' THEN PRINT 'and the lower limit on the MTTF is',1/LAMU
2000 PRINT
2010 PRINT RES$
2020 PRINT
2030 INPUT 'Do you want to analyse more data ',A$
2040 IF A$ = 'Y' THEN 300
2050 IF A$ = 'N' THEN STOP
2060 PRINT 'The response is Y(yes) or N(no). Try again.'
2070 GOTO 2030
2080 REM This section deals with binomial data.
2090 A3=1
2100 B3=1
2110 PRINT
2120 INPUT '                Do you have a prior',RP$
2130 IF RP$ = 'N' THEN GOTO 3050
2140 IF RP$ = 'Y' THEN GOTO 2170
2150 PRINT ' The response is Y(yes) or N(no). Try again.'
2160 GOTO 2120
2170 PRINT
2180 PRINT 'The program assumes the prior on the proportion'
2190 PRINT 'of failures is a bete distribution with parameters'
2200 PRINT 'alpha and beta. If you have your own values of'
2210 PRINT 'alpha and beta they can be inputs, otherwise they'
```

122 Estimating reliability

```
2220 PRINT 'can be calculated from prior information.'
2230 PRINT
2240 INPUT 'Do you have values for alpha and beta',RV$
2250 IF RV$ = 'Y' THEN GOTO 3000
2260 IF RV$ = 'N' THEN GOTO 2290
2270 PRINT ' The response is Y(yes) or N(no). Try again.'
2280 GOTO 2240
2290 REM This section calculates prior values of alpha and
2300 REM beta from prior information supplied by the operator
2310 PRINT
2320 PRINT 'It is assumed you know the most likely value of'
2330 PRINT 'p, the proportion of defective items, and the'
2340 PRINT 'upper 95% limit (i.e. the value p-max such that'
2350 PRINT 'there is a 95% probability that p is less than'
2360 PRINT 'p-max).'
2370 PRINT
2380 INPUT 'What is the most likely value',PM
2390 INPUT '      and the upper 95% limit',PU
2400 A1=1
2410 A2=2
2420 B2=(A2-1)*(1-PM)/PM+1
2430 TERM=(1-PU)**(B2+A2-1)
2440 SUM2=(1-TERM)/(B2+A2-1)
2450 FOR I=2 TO A2
2460 TERM=TERM*PU/(1-PU)
2470 SUM2=(SUM2*(I-1)-TERM)/(A2+B2-I)
2480 NEXT I
2490 SUM2=(A2+B2-1)*SUM2
2500 FOR I=1 TO A2-1
2510 SUM2=SUM2*(A2+B2-1-I)/I
2520 NEXT I
2530 IF SUM2 = 0 THEN GOTO 4680
2540 IF SUM2 > .95 THEN GOTO 2600
2550 A1=A2
2560 B1=(A1-1)*(1-PM)/PM+1
2570 SUM1=SUM2
2580 A2=A2+1
2590 GOTO 2420
2600 REM alpha lies between A1 and A2. Now find alpha by interpolation
2610 LAMU=PU
2620 D=.95-SUM1
2630 A3=A1+D*(A2-A1)/(SUM2-SUM1)
2640 B3=(A3-1)*(1-PM)/PM+1
2650 REM calculate beta(A3,B3)=gamma(A3)*gamma(B3)/gamma(A3+B3)
2660 NB=0
2670 X=A3+B3-1
2680 X=X-1
2690 NB=NB+1
2700 IF X>1 THEN GOTO 2680
2710 GOSUB 4320
2720 GAB=GX
2730 X=B3-1-NB
2740 GOSUB 4320
2750 GB=GX
2760 X=A3-1
2770 GOSUB 4320
2780 BAB=GB*GX/GAB
2790 FOR I=1 TO NB
2800 BAB=BAB*(B3-NB+I-1)/(B3+A3-NB+I-1)
2810 NEXT I
2820 REM Integrate using Simpson, and compare with the desired result
2830 GOSUB 3850
2840 IF ABS(SUM3-.95) < .0001 THEN GOTO 2920
2850 IF SUM3 < .95 THEN GOTO 2890
2860 SUM2=SUM3
2870 A2=A3
```

Further distributions   123

```
2880 GOTO 2620
2890 SUM1=SUM3
2900 A1=A3
2910 GOTO 2620
2920 PRINT
2930 PRINT RES$
2940 PRINT
2950 PRINT '       Prior values, Alpha equals',A3
2960 PRINT '                     Beta equals',B3
2970 PRINT
2980 PRINT RES$
2990 GOTO 3050
3000 REM This section accepts the known values of alpha and beta
3010 PRINT
3020 INPUT '      What is your value of Alpha',A3
3030 INPUT '                    and of Beta',B3
3040 GOTO 3050
3050 REM
3060 REM This section inputs binomial data for analysis
3070 REM
3080 PRINT
3090 PRINT 'You must now input the data you wish to analyse. The'
3100 PRINT 'data consists of the total sample size and the number'
3110 PRINT 'of defectives.'
3120 PRINT
3130 INPUT '              The sample size',NS
3140 INPUT 'and the number of defectives',ND
3150 IF 0 <= NS AND 0 <= ND AND ND <= NS THEN GOTO 3180
3160 PRINT 'This cannot be your data. Try again.'
3170 GOTO 3130
3180 A3=A3+ND
3190 B3=B3+NS-ND
3200 PRINT
3210 PRINT RES$
3220 PRINT
3230 PRINT '     Posterior values, Alpha',A3
3240 PRINT '                      Beta',B3
3250 PRINT
3260 PRINT RES$
3270 PRINT
3280 PM=(A3-1)/(A3+B3-2)
3290 PRINT
3300 PRINT RES$
3310 PRINT
3320 PRINT '  The most likely value of p is',PM
3330 REM calculate beta(A3,B3)=gamma(A3)*gamma(B3)/gamma(A3+B3)
3340 NB=0
3350 X=B3-1
3360 X=X-1
3370 NB=NB+1
3380 IF X>1 THEN GOTO 3360
3390 GOSUB 4320
3400 GB=GX
3410 X=A3+B3-1-NB
3420 GOSUB 4320
3430 GAB=GX
3440 X=A3-1
3450 GOSUB 4320
3460 BAB=GB*GX/GAB
3470 FOR I=1 TO NB
3480 BAB=BAB*(B3-NB+I-1)/(B3+A3-NB+I-1)
3490 NEXT I
3500 PU=PM
3510 NP=1
3520 LAMU=PU
3530 GOSUB 3850
```

124 Estimating reliability

```
3540 IF SUM3 > .95 THEN GOTO 3590
3550 NP=NP+1
3560 PU=PU+PM
3570 SUM1=SUM3
3580 GOTO 3520
3590 REM PU lies between (NP-1)*PM and NP*PM. Evaluate by linear interpolation.
3600 SUM2=SUM3
3610 P2=PU
3620 P1=PU-PM
3630 D=.95-SUM1
3640 PU=P1+D*(P2-P1)/(SUM2-SUM1)
3650 LAMU=PU
3660 GOSUB 3850
3670 IF ABS(SUM3-.95) < .0001 THEN GOTO 3750
3680 IF SUM3 < .95 THEN GOTO 3720
3690 P2=PU
3700 SUM2=SUM3
3710 GOTO 3630
3720 P1=PU
3730 SUM1=SUM3
3740 GOTO 3630
3750 PRINT
3760 PRINT 'The upper 95% Bayesian limit is',PU
3770 PRINT
3780 PRINT RES$
3790 PRINT
3800 INPUT 'Do you want to analyse more data ',A$
3810 IF A$ = 'Y' THEN 300
3820 IF A$ = 'N' THEN STOP
3830 PRINT 'The response is Y(yes) or N(no). Try again.'
3840 GOTO 3800
3850 REM
3860 REM The integration GOSUB. Integration done using Simpson's rule
3870 REM
3880 H=LAMU/10
3890 MX=5
3900 OI=0
3910 Y=LAMU
3920 GOSUB 4160
3930 LAST=FUN
3940 EVENS=0
3950 FOR I = 1 TO MX-1
3960 Y=2*I*H
3970 GOSUB 4160
3980 EVENS=EVENS+FUN
3990 NEXT I
4000 ODDS=0
4010 FOR I = 1 TO MX
4020 Y=(2*I-1)*H
4030 GOSUB 4160
4040 ODDS=ODDS+FUN
4050 NEXT I
4060 NI=(LAST+4*ODDS+2*EVENS)*H/3
4070 REM TEST FOR CONVERGANCE
4080 IF ABS(NI-OI) < .0001 THEN 4140
4090 EVENS=ODDS+EVENS
4100 OI=NI
4110 H=H/2
4120 MX=MX*2
4130 GOTO 4000
4140 SUM3=NI
4150 RETURN
4160 REM
4170 REM This GOSUB evaluates the function for use in the Simpson's
4180 REM rule GOSUB, at a point Y, and returned in the variable FUN.
4190 REM
```

```
4200 IF R$ = 'B' THEN GOTO 4290
4210 FUN=EXP(-Y*TH3)/GX
4220 FOR II=1 TO INT(A3-1)
4230 FUN=FUN*Y
4240 NEXT II
4250 DA=A3-INT(A3)
4260 FUN=FUN*(Y**DA)*TPA
4270 RETURN
4280 REM
4290 REM This section evaluates the Beta function.
4300 FUN=Y**(A3-1)*(1-Y)**(B3-1)/BAB
4310 RETURN
4320 REM
4330 REM This GOSUB evaluates the gamma function at the point X+1,
4340 REM using a polynomial approximation valid between zero and one.
4350 REM The value is returned in the variable GX
4360 REM
4370 GN=0
4380 IF X < 0 THEN GOTO 4510
4390 IF X < 1 THEN GOTO 4430
4400 X=X-1
4410 GN=GN+1
4420 IF X > 1 THEN GOTO 4400
4430 GOSUB 4610
4440 FOR GI=1 TO GN
4450 GX=GX*(X+GI)
4460 NEXT GI
4470 RETURN
4480 REM
4490 REM This section deals with the case when X is negative
4500 REM
4510 X=X+1
4520 GN=GN+1
4530 IF X < 1 THEN GOTO 4510
4540 GOSUB 4610
4550 FOR GI=0 TO GN-1
4560 GX=GX/(X-GI)
4570 NEXT GI
4580 RETURN
4590 REM
4600 REM This section calculates the polynomial.
4610 REM
4620 GX=GAMMA(1)
4630 FOR GI=2 TO 8
4640 GX=GX*X+GAMMA(GI)
4650 NEXT GI
4660 GX=GX*X+1
4670 RETURN
4680 PRINT 'SORRY, THIS PROBLEM IS INTRACTABLE FOR TECHNICAL REASONS'
```
Program 6.4 Bayesian data analysis performed on binomial, Poisson or exponential data

The data in the examples on confidence can be analysed using Bayesian techniques. As no prior is given, the reader should assume a 'flat prior'. This is assumed if the response 'N' is given in the program to the prompt 'Do you have a prior'.

Bayesian statistical analysis is not a subject that has been popular with authors of elementary texts, and for that reason most of the books on the subject go to some depth. A number of references are given in Chapter 2. Martz (Reference 11) gives a reasonable account,

126   Estimating reliability

particularly the applications to reliability. For the theory, Berger (Reference 3), Bain (Reference 1) and DeGroot (References 5 and 4 in order of increasing difficulty) are good. For more detail on the distributions mentioned in the book so far (not just in this section), the reader could do worse than read the above references, or Barlow (Reference 2), Gnedenko (Reference 7), Mann (Reference 10) or Lloyd (Reference 9).

6.8   Weibull analysis

Data that is believed to come from or be well fitted by a Weibull distribution is so often encountered in reliability that it was thought worthwhile giving it a section of its own. This is a classical analysis, but the problems associated with significance and confidence are not addressed here. It is possible to analyse Weibull data using Bayesian techniques, but it lacks the visual appeal of the graphical approach described below. Weibull himself was a Swedish engineer who investigated the distribution named after him in order to find a suitable description of the times to failure of mechanical components. Since then this distribution has proved so useful that it is possible to buy special graph paper (Weibull paper, described below) in order to ease the analytic process.

Suppose a number of items, such as pumps or bearings, are put on test, and run until they fail. The lifetimes of each item (i.e. the time from the start of the test, when they were new, until they fail), is recorded, and we wish to analyse this data. Suppose we tested n items, and the times to fail were

$$t_1, t_2, t_3, \ldots, t_n$$

We could assume that the times were exponentially distributed, in which case the best estimate of the MTTF is the mean of the n values $t_1, t_2, \ldots, t_n$, and the failure rate is the reciprocal of this. However it may be that we are in a situation where we know there is wear and hence some ageing process, and it is possible that the data could be Weibull distributed. It is possible to test this hypothesis statistically, though here we shall only describe a graphical method that underlies the analytic approach.

If the time to failure of an item is Weibull distributed, then the expression for the reliability is

$$R(t) = \exp - \left(\frac{t}{\gamma} \right)^{\beta}$$

where β is the shape parameter and γ is the characteristic life, or scale parameter. Take natural logs of both sides, to obtain

$$(t/\gamma)^\beta = -\ln(R)$$

and take logs again to give

$$\beta \ln(t) - \beta \ln(\gamma) = \ln(-\ln(R))$$

Rearrange this to get $\ln(t)$ in terms of $\ln(-\ln(R))$, giving

$$\ln(t) = \frac{1}{\beta} \ln(-\ln(R)) + \ln(\gamma)$$

Remember that R is the reliability, i.e. the proportion of items still functioning at time t, and this we can estimate (i.e. guess at) from the data. If the values of the t (the times to failure) are put into ascending order, then the following table shows the best estimate of R for each value of t.

t	t_1	t_2	t_3	$\ldots$	t_n
R	$1/(n+1)$	$2/(n+1)$	$3/n+1)$	$\ldots$	$n/(n+1)$

Notice that it is not n in the denominator of the fraction that gives R, but $n + 1$. This is for technical reasons that we cannot go into here in any detail, but is a generalization of the median. Suppose we only had three readings, the best guess at the median (the 50 % point) is the middle one. (The second one out of three gives the $50\% = 2/4$ point, and not the 2/3 point). If we now plot $\ln(t)$ against $\ln(-\ln(R))$, the result should be (approximately) a straight line with slope $1/\beta$ and intercept $\ln(\gamma)$. For a precise estimate, the straight line can be fitted by the method of least squares, which is the technique used in the computer program at the end of this section. For a more approximate estimate, the points can be plotted on graph paper, and the 'best' straight line fitted by eye, in which case the shape and scale parameters can be estimated from the graph, and an idea of the 'goodness of fit' of the data can be obtained.

Alternately, it is possible to buy 'Weibull' paper, on which the axes are marked with a log scale horizontally, and a log–log scale vertically. A number of variations on this scheme are available, all with appropriate techniques for estimating the shape and scale parameters. We shall describe only one of them here, with the help of an example. A device is tested to destruction by putting 15 of them on test. After 250 running hours, eight of them have been observed to fail. The failure times (in ascending order), and the estimates of the failure probabilities $(i/(n + 1))$ are given in the table below.

Time to failure (h)	27.5	42.5	102.5	138.0	148.0	165.0	189.0	249.0
Failure $i/(n+1)$ (%)	6.25	12.5	18.75	25.0	31.25	37.5	43.75	50.0

128 Estimating reliability

The analysis on Weibull paper is shown on Figure 6.11. The line AB is the 'best' line drawn by eye. The estimate of the scale parameter is obtained by drawing the line EP that is perpendicular to AB and passes through the 'estimation point' E. The estimate is obtained by noting where this passes through the scale marked $\hat{\beta}$, in this instance

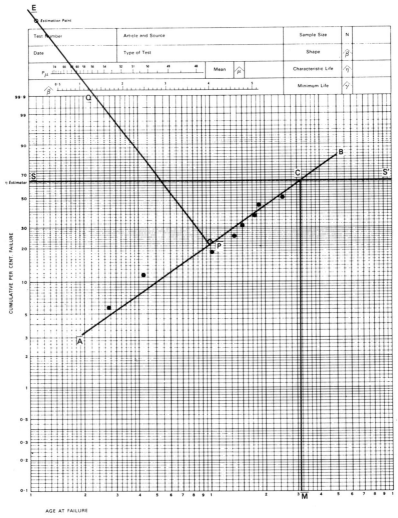

Figure 6.11

at the point Q at the value 1.2. The estimate of the shape parameter is obtained by observing that as

$$R = \exp - \left(\frac{t}{\gamma}\right)^{\beta}$$

then when

$$t = \gamma$$
$$R = 0.368$$

and so

$$F = 0.632$$

independent of the value of β. The line corresponding to the value 0.632 of F is drawn in and labelled estimate (the line SS'). The value of t corresponding to this value of F is along the line CM and in this case is about 315 h.

It is now possible to read other information off the graph, without having to do any calculations. For example, the reliability at time 30 h can be obtained by observing that

$$F = 5.5\%$$

when

$$t = 30 \text{ h}$$

and so the reliability at 30 h is 94.5% or 0.945. Similarly, the time at which 90% reliability is obtained is read off the graph, by noticing that when

$$F = 10\%$$

then

$$t = 50 \text{ h}$$

This technique is fine for obtaining approximate answers, but if more accurate figures are needed, then a calculation must be done, in which case,

$$R(30) = \exp - (30/300)^{1.2}$$
$$= 0.939$$

and

$$t = 300_{1.2}\sqrt{-\ln(90)}$$
$$= 45.99 \text{ h}$$

130 Estimating reliability

Program 6.5 does a least squares analysis on the data that is supplied by the operator, using the estimates of R discussed in the text. The program will also calculate R for a given value of time, and the time at which a stated reliability will be reached.

```
100 REM Weibull analysis
110 REM Data input
120 PRINT " How many items on test";
130 INPUT N
140 PRINT " and how many failed";
150 INPUT K
160 DIM T(K)
170 PRINT
180 PRINT " Enter the failure times"
190 FOR I=1 TO K
200 PRINT
210 PRINT " Failure time number ";I;
220 INPUT T(I)
230 NEXT I
240 REM Orders the failure times
250 REM from smallest to largest
260 FLAG=0
270 FOR I=2 TO K
280 IF T(I-1) <= T(I) THEN GOTO 330
290 TEMP=T(I)
300 T(I)=T(I-1)
310 T(I-1)=TEMP
320 FLAG=1
330 NEXT I
340 IF FLAG=1 THEN GOTO 260
350 REM Calculation of the parameters
360 XSUM,YSUM,XYSUM,X2SUM =0
370 FOR I=1 TO K
380 Y= LOG(T(I))
390 Z= 1/(1-(I/(N+1)))
400 X= LOG( LOG(Z))
410 XSUM=XSUM+X
420 YSUM=YSUM+Y
430 XYSUM=XYSUM + (X*Y)
440 X2SUM = X2SUM + (X*X)
450 NEXT I
460 XMEAN=XSUM/K
470 YMEAN=YSUM/K
480 TOP=(K*XYSUM)-(XSUM*YSUM)
490 BOT=(K*X2SUM)-(XSUM*XSUM)
500 B=BOT/TOP
510 REM *** B IS 1/BETA ***
520 C=YMEAN-(XMEAN/B)
530 THETA = EXP(C)
540 PRINT " Scale parameter";THETA
550 PRINT " Shape parameter";B
560 REM menu
570 PRINT
580 PRINT "                    Do you want the reliability (R)"
590 PRINT " or the time at which a reliability is achieved (T)"
600 PRINT "                    or to quit (Q)";
610 INPUT OPT$
620 PRINT
630 IF OPT$="R" OR OPT$="r" THEN GOSUB 680
```

```
640 IF OPT$="T" OR OPT$="t" THEN GOSUB 770
650 IF OPT$="Q" OR OPT$="q" THEN GOTO  820
660 GOTO 560
670 REM find the reliability
680 PRINT " Input the time";
690 INPUT CHRONOS
700 REM calculate the reliability
710 R = 1/EXP((CHRONOS/THETA)**B)
720 REM Convert to an integer %age
730 R = INT( R*100 + 0.5)
740 PRINT " The reliability at this time is ";R;" %"
750 RETURN
760 REM Find the time
770 INPUT " Input the %age reliability ";R
780 R=R/100
790 CHRONOS = THETA*(LOG(1/R))**(1/B)
800 PRINT " Reliability achieved at time";CHRONOS
810 RETURN
820 END
```

Program 6.5 Weibull analysis

The technique of estimating Weibull parameters is very well documented and several references are given in Chapter 2. O'Connor (Reference 12) is very clear on this, and Def. Stan. 00-41 (Reference 18) gives an account as well, and includes a technique for putting confidence limits on the parameters. Mann (Reference 10) goes into other ways of dealing with Weibull data in some detail, and is well worth reading.

WORKED EXAMPLES

(6.1) One defective is observed in a random sample of five items when the proportion of defective items is believed to be no worse than one in 200. Is this result significant?

The null hypothesis is that the proportion of defectives is no greater than 0.5%. In that case

$$P \text{ (0 defectives)} = (0.995)^5$$

$$= 0.975$$

$$P \text{ (1 or more defectives)} = 1 - 0.975$$

$$= 2.5\%$$

This is less than 5%, so the result is significant at the 5% level of significance, but greater than 1%, and so is not significant at the 1% level of significance. It is not a great cause for alarm.

(6.2) Consider the case when five items are examined, with a

132 Estimating reliability

probability 0.05 of a single one being defective. Then if r is the number of defectives,

r	p_r		
0	0.95^5	$= 0.774$ (note that $0! = 1$ and $C_n^n = 1$)	
1	$5 \times 0.05 \times (0.95)^4$	$= 0.204$	
2	$\dfrac{5 \times 4}{1 \times 2} \times (0.05)^2 \times (0.95)^3$	$= 0.021$	
3	$\dfrac{5 \times 4 \times 3}{1 \times 2 \times 3} \times (0.05)^3 \times (0.95)^2$	$= 0.001$	
4	$\dfrac{5 \times 4 \times 3 \times 2}{1 \times 2 \times 3 \times 4} \times (0.05)^4 \times 0.95$	$= 0.00003$	
5	$(0.05)^5$	$= 0.0000003$	

Notice that the total is one (subject to rounding errors). In real terms, this means that if we were to examine such a sample every day, then on about three days in every four (on average) we would find no defectives, on one day in five, one defective, on one day in 50, two defectives, and one day in a thousand (i.e. once every several years), three defectives, and four or five defectives would be found so rarely as to be discounted. In practice, if four or five defectives were found, then the 'only 5% defective' statement would be extremely suspect. This is the basis of our significance test! In practice it is not necessary to make a full table, as the following examples show.

(6.3) A manufacturer claims that the quality of his goods is better than 98% (i.e. no more than 2% of them are defective). A sample of ten items chosen at random has two defective items. Is this result significant?

The null hypothesis is that p (defective) is no more than 2%. If this hypothesis is correct,

$$P \text{ (0 or 1 defective)} = p_0 + p_1$$
$$= (0.98)^{10} + 10(0.98)^9 \times 0.02 \text{ using the binomial distribution}$$
$$= 0.984$$
$$P \text{ (2 or more defectives)} = 1 - 0.984$$
$$= 0.016$$

Worked examples 133

This is less than 0.05, so the result is significant at the 5% level, but greater than 0.01, so the result is not significant at the 1% level. In subjective terms, it would be considered cause for concern!

(6.4) A manufacturer produces goods to a stated quality of 1% defective. In a sample of 50 items, three are defective. Is this cause for concern?

The null hypothesis is that $p = 0.01$. Then

P (3 or more defectives)

$$= 1 - P \text{ (2 or fewer defectives)}$$

P (2 or fewer defectives)

$$= p_0 + p_1 + p_2$$

$$= (0.99)^{50} + 50 \times 0.01 \times (0.99)^{49} + \frac{50 \times 49}{2}$$

$$\times (0.01)^2 \times (0.99)^{48}$$

$$= 0.986$$

P (3 or more defectives)

$$= 1 - 0.986$$

$$= 1.4\%$$

This is significant at the 5% level, but not at the 1% level.

(6.5) When my car is well tuned, I know from experience that I may have difficulty starting it on average about once a month. During one particularly bad week, I experienced three troublesome instances when it is difficult to start. Does this indicate it is due for a service?

In a month the average number of poor starts is 1, so in a week it is about 0.25. This is the null hypothesis — that $m = 0.25$.

$$P \text{ (3 or more poor starts)} = 1 - P \text{ (2 or fewer poor starts)}$$

$$P \text{ (2 or fewer poor starts)} = p_0 + p_1 + p_2$$

$$= e^{-m} + m \, e^{-m} + \frac{m^2}{2} e^{-m}$$

$$= 0.993$$

$$P \text{ (3 or more poor starts)} = 1 - 0.998$$

$$= 0.2\%$$

This is highly significant, so I conclude that m is a great deal larger than 0.25 — or it's about time my car was serviced!

134 Estimating reliability

(6.6) The probability of an item being defective is 0.05. In a sample of ten such items, calculate the probabilities of $0, 1, 2, \ldots$ defectives.

Using the binomial distribution with $n = 10$ and $p = 0.05$, and comparing the results with the Poisson distribution when $m = np = 0.5$, we get the following table:

r	Binomial	Poisson
0	0.599	0.607
1	0.315	0.303
2	0.075	0.076
3	0.010	0.013
4	0.001	0.002

It can be seen that there is a reasonable agreement between the two columns, particularly when the probabilities are of a reasonable magnitude. The agreement is better for larger values of n and small values of p.

(6.7) The failure rate of my central heating system is 0.5 failures/year. It is serviced at the beginning of the autumn each year. What is the probability of it failing during Christmas week when all the mechanics are on holiday?

Measure time in weeks from October 1st. Then $\lambda = 1/104$ failures/week. Christmas week is between $t = 12$ and $t = 13$.

$$P(12 < t < 13) = \int_{12}^{13} \lambda e^{-\lambda t} \, dt$$

$$= \left(-e - \frac{t}{104} \right)_{12}^{13}$$

$$= 0.01 \text{ or } 1\%$$

It'll never happen!

(6.8) One defective item in a sample of 20 is observed. What is the 99% confidence limit on p, the proportion of defectives?

If p satisfies

$$q^{20} + 20q^{19}p > 0.01$$

then the number of defectives will not be significantly low. The solution of this inequality is

$$p < 0.29$$

so we are 99% confident that p is less than 0.29, which is the 99% confidence limit on p.

Worked examples 135

(6.9) In a year, two breakdowns are observed in the family washing machine. What is the 90 % confidence limit on m, the average number of breakdowns that may be expected in a year?

Using the Poisson distribution for the number of breakdowns observed in a year, as few as two would be reasonable (i.e. not significant at the 10 % level) if

$$e^{-m} + m\,e^{-m} + \frac{m^2}{2}\,e^{-m} < 0.1$$

The solution of

$$e^{-m} + m\,e^{-m} + \frac{m^2}{2}\,e^{-m} = 0.1$$

is

$$m = 5.32$$

If m were larger than this value, then as few as two breakdowns would be significant at the 10 % level. The 90 % confidence limit on m is 5.32.

(6.10) Six components were tested to failure. The mean of the observed failure times was 77 h. What is the 95 % confidence limit on λ, the failure rate of these components (assumed constant), and on the MTTF and the reliability at 10 h?

The distribution of the sum of the failure times, T, is an Erlang distribution:

$$F(T) = \frac{\lambda^7}{7!}\,T^6\,e^{-\lambda T}$$

Then if

$$\int_{462}^{\infty} F(T)\,\mathrm{d}T > 0.05$$

the result is not significantly high at the 95 % level of significance. The solution of

$$\int_{462}^{\infty} F(T)\,\mathrm{d}T = 0.05$$

is

$$\lambda = 0.023$$

If λ is greater than this value, the observed MTTF of 77 h (or T of 462) would be significant. Then 0.023 is the confidence limit on λ. The lower confidence limit on the MTTF, M_1, is given by

$$M_1 = 1/\lambda$$
$$= 43.9$$

136 Estimating reliability

and the lower confidence limit on the reliability at 10 h, R_l, is given by

$$R_l = \exp - 10/43.9$$
$$= 0.796$$

(6.11) An electronic equipment is tested each week by means of a hand-operated test unit. The manufacturer of the test unit claims that the reliability of the unit is 99 %, while the false alarm rate is only 2 %. What conclusions can be drawn?

Let F be the event 'the equipment is faulty' and D the event 'the test unit shows a fault'. Then presumably the manufacturer's claims are that

$$P(D|F) = 0.99$$

i.e. the probability of the unit showing a fault if one is present is 99 %, and

$$P(D|\bar{F}) = 0.02$$

where the bar denotes negation, i.e. the probability of the unit showing a failure if one is not present is 2 %.

What is of interest to the owner of the equipment is

$$P(F|D)$$

and

$$P(F|\bar{D})$$

At this stage it might be profitable to ask why we assume this interpretation of the manufacturer's claims. The answer is twofold:

1. the former data is (relatively) easy to collect, and
2. the second pair of probabilities is related to the first through Bayes' theorem, and to use Bayes' theorem $\pi(F)$, the prior probability of a fault being present, is required. The manufacturer cannot possibly know $\pi(F)$ because it depends on equipment usage and a number of other things.

Put

$$\pi = \pi(F)$$

Then, by Bayes' theorem,

$$P(F|D) \propto P(D|F)\pi(F) = 0.99\pi$$

and

$$P(\bar{F}|D) \propto P(D|\bar{F})\pi(\bar{F}) = 0.02(1 - \pi)$$

and as these probabilities must sum to one,

$$P(F|D) = \frac{0.99}{0.99\pi + 0.02(1 - \pi)}$$

If faults are rare, say $\pi = 0.01$, then

$$P(F|D) = 0.33$$

i.e. of every three faults occurring, two are missed.
If faults are less rare, say $\pi = 0.1$, then

$$P(F|D) = 0.85$$

(6.12) It is necessary to estimate the number of breakdowns of a fleet of lorries in a week. The number is Poisson distributed, with mean m, and the fleet manager believes the most likely value of m is 1.5, but he is 95% certain that it is below 2.5. Using Program 6.4, this gives a gamma prior with shape parameter 11.7 and scale parameter 7.1, i.e.

$$\pi(m) = \frac{(7.1)^{12.7}}{\Gamma(11.7)} \lambda^{11.7} e^{-7.1\lambda}$$

In a given week, two breakdowns are observed. The probability of this happening is the likelihood function (as there is only one observation), and so

$$L = e^{-m} \frac{m^2}{2}$$

This gives

$$P(m) = \frac{(8.1)^{14.7}}{\Gamma(13.7)} \lambda^{13.7} e^{-8.1\lambda}$$

i.e. the posterior is gamma with shape parameter 13.7 and scale parameter 8.1. This gives

$$\text{MLE}(m) = 1.7$$

and the 95% Bayesian upper limit

$$m_\mu = 2.5$$

PROBLEMS

(6.1) Test, at the 5% level, the significance of the result of testing to failure, under conditions of high stress, five electronic circuits that

138   Estimating reliability

failed at times (in hours)

$$1283 \qquad 1895 \qquad 2021 \qquad 2487 \qquad 2510$$

if the hypothesis is that the MTTF is no more than 2000 h.
(*Solution:* the probability is 5.8%, the result is not significant.)

(6.2) A plant manager observes five major breakdowns of the plant in one particular month, when he believes the average number of breakdowns is two. Is this result significant?
(*Solution:* the probability is 5.2%, not significant.)

(6.3) A manufacturer of light bulbs believes there is a 95% chance of his product continuing to function during a particular test cycle. He tests a sample of 200, and 182 survive. Is this result significant?
(*Solution:* the probability is 1.2%, significant at the 5% level, but not at the 1%.)

(6.4) Eight cooling fans from a copier are tested until they fail. A duty cycle consists of the fan being switched on, left running for 15 seconds and then being switched off. The number of duty cycles until failure of each of the eight fans was, in thousands of duty cycles,

$$12.2 \qquad 13.8 \qquad 14.9 \qquad 15.4 \qquad 15.5 \qquad 17.9 \qquad 19.2 \qquad 22.8$$

Find the 95% upper confidence limit on λ, the failure rate in failure per thousand duty cycles, and the lower confidence limit on the mean number of cycles to failure. (The 95% upper confidence limit on λ is 0.1 failures/thousand cycles, and the 95% lower confidence limit is 9941 cycles.)

(6.5) In a fleet of cars it was observed that on one particular day six cars in the fleet were out of action. Assuming the number of such cars is Poisson distributed, put a 90% upper confidence limit on the mean number of cars out of action each day.
(*Solution:* 10.5.)

(6.6) A manufacturer of self-heating cans of food tested a random sample of his product. Of 100 cans, only one of them failed to function properly. Put a 99% lower confidence limit on the proportion of cans that will function.
(*Solution:* 0.935.)

(6.7) Using the data of the worked example 6.11, calculate the probability of a false alarm, i.e. calculate $P(F|\bar{D})$.
(*Solution:* $0.01/(0.01\pi + 0.93(1 - \pi)) = 0.0001$ if $\pi = 0.01$ and 0.001 if $\pi = 0.1$.)

Problems 139

(6.8) In order to decrease the amount of time chasing non-existent faults, the test unit must obviously be improved. In what way? Study the effect of increasing the reliability to 99.5%, and then separately, the effect of improving the false alarm rate to 1%, in the case when π is 0.01.

(*Solution:* the values of $P(F|D)$ are 0.334 and 0.5 respectively.)

(*Note:* The manufacturer's claims about his equipment appear very good initially, but in order to assess its effectiveness, the failure rate of the equipment under test must be known and appear in the equation, when, in this case, the test equipment does not appear in such a good light. As an exercise, using 0.01 as the probability of equipment failure, the reader might like to examine how good the manufacturer's test unit should be if limits are put on $P(F|D)$ and $P(F|\bar{D})$, i.e. if

$$P(F|D) > 0.99$$

and

$$P(F|\bar{D}) < 0.02$$

what values should $P(D|F)$ and $P(D|\bar{F})$ take?)

(6.9) The owner of a car hire firm observed the following about the cylinder-head gaskets of one of the makes of car he owned. All the cars were purchased new, and when one of the gaskets failed, it was replaced with a new one. The following failure distances, in miles, were observed for six gaskets:

33 289	64 218	78 521	86 483	92 917	98 103

while seven gaskets were still functioning after the following distances:

1382	10 219	12 284	30 193	60 128	65 917	94 803

All the gaskets were produced by the same company, and used under similar conditions. The manufacturer believed the most likely value for the MDTF was 100 000 miles, and that it was at least 65 000 miles; (the 95% Bayesian lower limit). Using the prior information, and the data given, calculate the posterior MLE of the MDTF and the 95% Bayesian lower limit.

(*Solution:* 101 288 miles and 73 717 miles. NB: Because of the censored data, this problem would be much, much more difficult using classical analysis.)

(6.10) The safety manager of a factory believes that the severe accident rate at his factory is 1 per year. In any case, it is no more than 1.5 per year. In one particular year, two accidents are observed. How

140 Estimating reliability

must this affect his belief if he uses Bayesian analysis to deal with this data? If the following year there are no accidents, how does this further effect his belief?

(*Solution:* mean 1.11, upper 95% limit 1.55, mean 1.105, upper limit 1.53.)

(**6.11**) A manufacturer of inflatable dingies tests them by overinflating them to a standard pressure. It is believed that 99% of them will pass this test without rupturing, and in any case n more than 2% will fail. On testing a sample of 20, one fails. How should this effect his prior?

(*Solution:* MLE of proportion of failures is 1.1%, 95% upper limit is 2.2%.)

(**6.12**) Thirty long-life items were tested for 100 hours, by which time 6 of them had failed at times (in hours):

68   74   76   89   95   99

Assuming the lifetimes are Weibull distributed, find an estimate of the scale and shape parameters, the median life (i.e. the time at which half of the items will be expected to fail), and the time at which 5% will have failed.

Perform your analysis using Program 6.5. Also plot the data on Weibull paper, and compare the results.

(*Solution:* Scale = 147 h, shape = 4.1, $R(134) = 0.5$, $R(71.2) = 0.95$.)

Chapter 7

Quality assurance

7.1 Introduction

Once the equipment is designed, developed, tested and the final design is accepted, it is handed over to the production team for manufacture. At this stage care must be taken to ensure that the item produced is the one that the designers have created, and furthermore that it is produced over and over again. It is at this stage that the quality control team takes over to ensure the quality and reliability of the product. However, it must be emphasized that reliability cannot be inspected into a product — it must be designed in from the start. The reason for inspection is to ensure that the manufactured quality is that which the designer intended.

Ideally every item should be inspected. This, however, is not always possible, particularly if the inspection involves a destructive test (as it must with pyrotechnics and explosives, for example). In this case, a random sample of the items under consideration are tested, and on the basis of the results of that inspection, they are all accepted as fit for use, or they are rejected.

The situation we shall examine here is the one in which a customer is purchasing a batch, or lot, of items from a manufacturer. The batch may consist of several thousand items, or only a small number. The customer is going to examine a random sample from the batch, and each item in the sample will be classified as either acceptable or nonacceptable (good or bad, with no degraded state). This is called 'inspection by attribute'. On the basis of this inspection, he will either accept the batch as being fit for use, and accept the consequences if he had made a mistake, or he will reject it, in which case the manufacturer will suffer any penalties that may arise from possible error.

The accept/reject criterion is on the lines of

'Reject if there are too many nonacceptoable items, otherwise accept.'

The sample size and the acceptable number of dud items in the sample has to be agreed by both the customer and supplier. The

142 Quality assurance

problem arises in determining those numbers because with any scheme there is a degree of risk attached for both customer and supplier. The example which follows illustrates this point.

A demolition company is purchasing blast caps from a manufacturer. The only way of determining the quality of a cap is to detonate it, in which case it cannot be reused. The company decide to test 20 from a large batch, and if two or more were to fail to function satisfactorily, the company will reject the batch, while if one or none fail, the batch will be accepted.

It is assumed that a proportion p (unknown) of the batch is not acceptable, that the sample is taken at random, and that the batch is so large that the number of duds in the sample is binomially distributed. Then the probability of accepting the batch, as a function of p, is

$$P\text{ (Accepting)} = p_0^1 + p_1$$
$$= p^{20} + 20p^{19}q \qquad \text{where } p = 1 - q$$

The graph of this function, which is known as the operating characteristic, or OC, is shown in Figure 7.1.

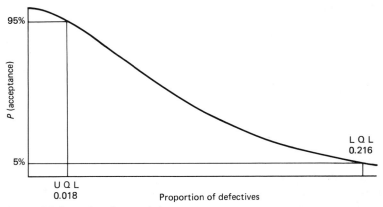

Figure 7.1 Operating characteristic

The value of p of 1.8% is such that P (Acceptance) is 95% (and hence that P (Rejection) is 5%) is known as the upper quality level (UQL) and is an indication of the producer's risk (that a good batch is condemned as being unacceptable), while the value of p of 21.6% which corresponds to the value of 5% for $P(A)$ is called the lower quality level (LQL), and is an indication of the customer's risk. Both these points are shown on Figure 7.1. Quality control schemes such as this one are often chosen on the basis of agreed producers and customers risks. The values chosen will reflect the relative costs of making a mistake, also taking into account the cost of inspection.

In a second example a shipping company wishes to purchase flares for use if a ship is in distress. It sets a 10% LQL at 0.116. The manufacturer sets a 95% UQL at 0.0113. What sample scheme should be used?

The statements on the LQL and the UQL must be interpreted as

$$P(\text{Acceptance at } p = 0.0113) > 0.95$$

$$P(\text{Acceptance at } p = 0.116) < 0.10$$

The unknowns in the equation are n, the sample size, and r, the maximum number of acceptable duds in the sample. The distribution of the number of duds in the sample is assumed binomial, then n and r must satisfy

$$p_0 + p_1 + \cdots + p_r > 0.95 \quad \text{when } p = 0.0113$$

$$p_0 + p_1 + \cdots + p_r < 0.10 \quad \text{when } p = 0.116$$

The Program 7.1 solves these inequalities, the solutions of which are that a sample of 52 must be taken, reject if two or more defectives are seen, accept otherwise, and the OC is shown below in Figure 7.2.

Notice that the curve must pass above the UQL, but below the LQL.

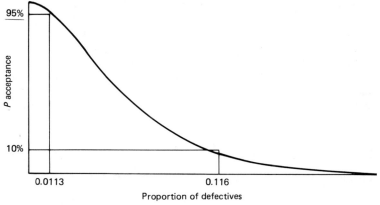

Figure 7.2 Operating characteristic

There are still questions that can be asked concerning such schemes. What, for example, can be said about the quality of goods that the customer eventually receives, on average, if he is buying several batches of identical items, and is testing each batch? What happens to rejected batches?

The customer is not too keen to see batches rejected. After all, he is in the business of buying goods from the supplier, and putting them to use himself. He is interested in receiving goods of as high a quality as possible. If a rejected batch could be 100% inspected, and corrected, so that it contained no dud items, the situation shown in Figure 7.3 would pertain. This shows the average quality received by

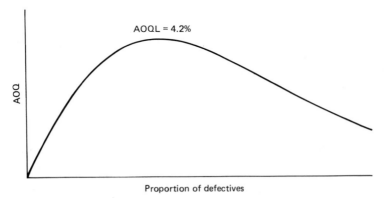

Figure 7.3 Average outgoing quality

the customer (and it is only an average) as a function of p, received by the customer over a long period of receiving batches from the supplier, for the scheme discussed in the first example (the demolition company). This function is called the average outgoing quality. Notice that this has a maximum at the value of 4.2%. This is the worst average value the customer will receive using this scheme, and is called the average outgoing quality limit (AOQL). So for example the AOQL for the second example is 2.6%, and the graph is shown in Figure 7.4.

7.2 Multiple sampling schemes

Consider the following scheme, which is a bit more complex than any considered so far.

Examine thirteen items. Reject if two or more are dud, and accept if none are dud. Otherwise, examine again, and this time reject if the total number of duds is two or more (that is the sum of the number of duds from the first and second sample). What are the advantages of such a scheme?

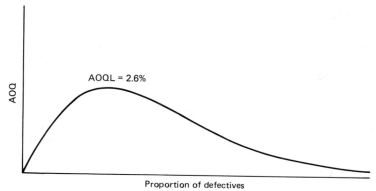

Figure 7.4 Average outgoing quality

Clearly, if the batch is either very good or very bad, only one sample will be examined (as the reject/accept decision will be made on one sample). It is only in the case of the sample falling into the 'grey area' that two samples need be examined. The probability of acceptance is

$P(A) = P$ (acceptance on the first sample, or doubt on the first sample and acceptance on the second)

$= P_0 + P_1(P_0 + P_1)$

while the OC and AOQ curves are shown in Figure 7.5.

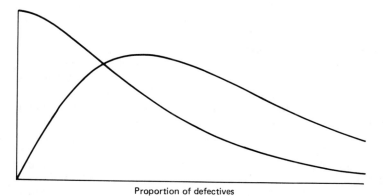

Figure 7.5 AOQ and OC curves for double sampling scheme

The UQL is 1.8%, and the LQL is 21.6%. This is very similar to the first example. So why consider such a scheme? The advantage lies in the relative economy of the scheme. If the batch is either very good or very poor, then almost certainly only one sample is examined. It is only if the quality is indifferent that there is a reasonable chance that two batches are examined. Figure 7.6 shows the graph of the average number of items inspected (and again it is only an average figure over several submissions) as a function of submitted quality. The horizontal line shows the number of items examined for the corresponding single sampling scheme. Readers will be able to see the savings for themselves.

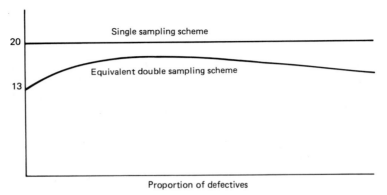

Figure 7.6 Average number of items inspected

It was not the purpose of this chapter to give more than a brief look at the ideas behind quality assurance. For details of the plans that are available see the references in Chapter 2. BS 6001 (Reference 16) is available, and BS 5750 (Reference 14) describes QA. O'Connor (Reference 12) gives a good exposition on the use of BS 6001.

There are two programs for this section. Program 7.1 calculates a simple sampling scheme for a given UQL and LQL, and calculates the AOQL, by finding the maximum value of the AOQ. Program 7.2, which is somewhat longer, performs a variety of tasks, and is menu driven. The scheme is input by the operator, single or double, and the operator has the choice of a point on the OC curve, the AOQ and average number inspected if the scheme is a double scheme, for a stated value of p, the presented quality; the value of p corresponding to a stated point on the OC curve, hence allowing the UQL and LQL to be obtained; the AOQL; and in the case of a double sampling scheme, the maximum value of the average number inspected.

Multiple sampling schemes   147

The programs are not complex, though Program 7.2 is rather long. The only complication is the quadratic approximation used to find the appropriate maximum. Three points are chosen, and a quadratic is drawn through the three points on the curve. The maximum of the quadratic is found, and this replaces the point at which the value on the curve is least. The process is then repeated with these three points until the values of the function are sufficiently close together, at which point this must be at, or very close to, the maximum.

```
100 REM
110 REM
120 REM
130 PRINT ' This routine calculates sampling plans for'
140 PRINT ' a given LQL and UQL. The inputs are the'
150 PRINT ' probabilities of acceptance at the LQL and'
160 PRINT ' UQL and the quality in terms of the proportion'
170 PRINT ' of defectives. The former values are usually'
180 PRINT ' .05 and .95, but the first two input give'
190 PRINT ' the user the chance to input his own values'
200 PRINT
210 PRINT ' Input the P(A) at the LQL';
220 INPUT PL
230 PRINT '          and P(A) at the UQL';
240 INPUT PU
250 IF 0<PL AND 0<PU AND PU<1 AND PL<1 AND PL<PU THEN GOTO 280
260 PRINT ' Your input is out of range. Try again.'
270 GOTO 210
280 PRINT '              Input the LQL';
290 INPUT LQL
300 PRINT '                  and the UQL';
310 INPUT UQL
320 IF 0<LQL AND 0<UQL AND LQL<1 AND UQL<1 AND LQL>UQL THEN GOTO 35C
330 PRINT ' Your input is out of range. Try again'
340 GOTO 280
350 REM Try acceptance of 1 first
360 A=1
370 N=INT(LOG(PL)/LOG(1-LQL))+1
380 IF (1-UQL)**N >= PU THEN GOTO 620
390 NT=N
400 TERM=1
410 SUM=0
420 FOR I=0 TO A
430 SUM=SUM+TERM*(1-LQL)**(NT-I)*LQL**I
440 TERM=TERM*(NT-I)/(I+1)
450 NEXT I
460 IF SUM<=PL THEN GOTO 490
470 NT=NT+1
480 GOTO 400
490 REM NT,A is the scheme that fits at the LQL.
500 REM Test at the UQL
510 SUM=0
520 TERM=1
530 FOR I=0 TO A
540 SUM=SUM+TERM*(1-UQL)**(NT-I)*UQL**I
550 TERM=TERM*(NT-I)/(I+1)
560 NEXT I
570 IF SUM >= PU THEN GOTO 620
580 A=A+1
```

148 Quality assurance

```
590 N=NT
600 GOTO 400
610 REM N,A-1 is the scheme required
620 PRINT
630 PRINT ' The required scheme is:'
 640 PRINT '                Sample size is';N
650 PRINT '    with an acceptance number of';A-1
660 REM Now to find the AOQL for the scheme
670 REM Uses a quadratic approximation
680 P1=UQL
690 SUM1=0
700 TERM=1
710 FOR I=0 TO A-1
720 SUM1=SUM1+TERM*(1-P1)**(N-I)*P1**I
730 TERM=TERM*(N-I)/(I+1)
740 NEXT I
750 SUM1=SUM1*P1
760 P2=LQL
770 SUM2=0
780 TERM=1
790 FOR I=0 TO A-1
800 SUM2=SUM2+TERM*(1-P2)**(N-I)*P2**I
810 TERM=TERM*(N-I)/(I+1)
820 NEXT I
830 SUM2=SUM2*P2
840 P0=(P1+P2)/2
850 SUM0=0
860 TERM=1
870 FOR I=0 TO A-1
880 SUM0=SUM0+TERM*(1-P0)**(N-I)*P0**I
890 TERM=TERM*(N-I)/(I+1)
900 NEXT I
910 SUM0=SUM0*P0
920 REM order according to the values of SUM*
930 REM put largest in SUM0 and smallest in SUM2
940 IF SUM0>SUM1 THEN GOTO 1010
950 SUM=SUM0
960 SUM0=SUM1
970 SUM1=SUM
980 P=P0
990 P0=P1
1000 P1=P
1010 IF SUM0>=SUM2 THEN GOTO 1080
1020 SUM=SUM0
1030 SUM0=SUM2
1040 SUM2=SUM
1050 P=P0
1060 P0=P2
1070 P2=P
1080 IF SUM1>=SUM2 THEN GOTO 1150
1090 SUM=SUM1
1100 SUM1=SUM2
1110 SUM2=SUM
1120 P=P1
1130 P1=P2
1140 P2=P
1150 REM if difference max to min < .001 then stop
1160 IF SUM0-SUM2 <=.001 THEN 1310
1170 D0=(SUM1-SUM2)*P0
1180 D1=(SUM2-SUM0)*P1
```

Multiple sampling schemes   149

```
1190 D2=(SUM0-SUM1)*P2
1200 NM=P0*D0+P1*D1+P2*D2
1210 DEN=D0+D1+D2
1220 P2=NM/DEN/2
1230 SUM2=0
1240 TERM=1
1250 FOR I=0 TO A-1
1260 SUM2=SUM2+TERM*(1-P2)**(N-I)*P2**I
1270 TERM=TERM*(N-I)/(I+1)
1280 NEXT I
1290 SUM2=SUM2*P2
1300 GOTO 920
1310 PRINT
1320 PRINT '                         The  AOQL is';SUM0
1330 PRINT
1340 PRINT ' Do you want to look at another scheme';
1350 INPUT R$
1360 IF R$='Y' THEN GOTO 200
1370 IF R$='N' THEN GOTO 1400
1380 PRINT ' Input is Y(yes) or N(no).Try again!'
1390 GOTO 1340
1400 END
```

Program 7.1 Simple sampling scheme for a given LQL and UQL

```
100 REM
110 REM
120 REM
130 PRINT ' This routine analyses sampling plans. The input is the'
140 PRINT ' plan (single or double, and the output can be the LQL,'
150 PRINT ' the UQL, the AOQL, the average number inspected, the AOQ'
160 PRINT ' or the probability of accptance. What sort of scheme?'
170 PRINT '                         Single(S) or double(D)';
180 INPUT S$
190 IF S$='S' OR S$='D' THEN GOTO 220
200 PRINT ' The response is S (single) or D (double)'
210 GOTO 170
220 PRINT
230 PRINT ' Input the parameters of the scheme. The sample size first'
240 PRINT ' (if a double scheme, it is assumed the two samples are the'
250 PRINT '                         same size)';
260 INPUT N
270 PRINT
280 IF S$='D' THEN GOTO 370
290 PRINT ' Now the acceptance number-that is the maximum number'
300 PRINT '             of defectives acceptable in the sample';
310 INPUT A
320 IF N=INT(N) AND N>0 AND A=INT(A) AND A>=0 AND A<N THEN GOTO 350
330 PRINT ' Input error! Try again.'
340 GOTO 230
350 PRINT
360 GOTO 580
370 PRINT ' For a double scheme, the acceptance number for the first'
380 PRINT ' sample is needed-that is the maximum number of defectives'
390 PRINT '                         acceptable';
400 INPUT A
410 PRINT
420 PRINT ' and the rejection number for the first sample-that is,'
430 PRINT ' the minimum number of defectives needed to reject on the'
440 PRINT '                         first sample';
450 INPUT R
```

150 Quality assurance

```
460 PRINT
470 PRINT ' and the acceptance number for the second sample-that is,'
480 PRINT ' the maximum number of acceptable defectives in the two'
490 PRINT '                                          samples together';
500 INPUT A2
510 IF N<>INT(N) OR A<>INT(A) OR A2<>INT(A2) OR R<>INT(R) THEN GOTO 530
520 IF N>0 AND A>=0 AND A<R AND R<N AND A2>=R-1 THEN GOTO 550
530 PRINT ' Input error! Try again.'
540 GOTO 370
550 PRINT
560 ND=R-A
570 DIM PD(ND)
580 PRINT
590 PRINT ' ****************************************************************'
600 PRINT ' *                         MENU                              *'
610 PRINT ' *                                                           *'
620 PRINT ' *   Type 1 for P(A) and other functions                    *'
630 PRINT ' *   Type 2 for the AOQL                                     *'
640 P$= ' *   Type 3 for the max value of the average no. inspected *'
650 IF S$ = 'D' THEN PRINT P$
660 PRINT ' *   Type 4 for P as a function of P(A)                     *'
670 PRINT ' *   Type 5 to input another scheme                         *'
680 PRINT ' *   Type 6 to exit the program                             *'
690 PRINT ' *                                                           *'
700 PRINT ' ****************************************************************'
710 PRINT
720 PRINT '                                                      Choice';
730 INPUT D$
740 IF D$='1' OR D$='2' OR D$='4' OR D$='5' OR D$='6' THEN GOTO 780
750 IF S$='D' AND D$='3' THEN GOTO 780
760 PRINT ' Input error! Try again.'
770 GOTO 580
780 IF D$='1' THEN GOSUB 850
790 IF D$='2' THEN GOSUB 1340
800 IF D$='3' THEN GOSUB 1850
810 IF D$='4' THEN GOSUB 2370
820 IF D$='5' THEN GOTO 170
830 IF D$='6' THEN GOTO 2740
840 GOTO 580
850 REM This section calculates the P(A) etc
860 PRINT '         Input a value of P, the proportion of defectives';
870 INPUT P
880 IF 0<=P AND P<=1 THEN GOTO 910
890 PRINT ' Input error! Try again.'
900 GOTO 860
910 GOSUB 990
920 IF S$='D' THEN GOSUB 1230
930 PRINT
940 PRINT '                        The probability of acceptance is';PA
950 PRINT '                                          The AOQ is';P*PA
960 DAV$= '                        The average number inspected is'
970 IF S$='D' THEN PRINT DAV$;AV
980 RETURN
990 REM This GOSUB calculates P(Acceptance)
1000 REM The input is P, the proportion of defectives.
1010 REM The output is PA.
1020 PA=0
1030 TERM=1
1040 FOR I=0 TO A
1050 PA=PA+TERM*(1-P)**(N-I)*P**I
1060 TERM=TERM*(N-I)/(I+1)
```

Multiple sampling schemes 151

```
1070 NEXT I
1080 IF S$='S' THEN RETURN
1090 FOR I=A+1 TO R-1
1100 PD(I-A)=TERM*(1-P)**(N-I)*P**I
1110 TERM=TERM*(N-I)/(I+1)
1120 NEXT I
1130 REM
1140 PA2=0
1150 TERM=1
1160 FOR I=0 TO A2-A-1
1170 PA2=PA2+TERM*(1-P)**(N-I)*P**I
1180 TERM=TERM*(N-I)/(I+1)
1190 IF I<A2-R+1 THEN GOTO 1210
1200 PA=PA+PA2*PD(A2-A-I)
1210 NEXT I
1220 RETURN
1230 REM This GOSUB calculates the average number inspected
1240 REM The input is P, the proportion of defectives
1250 REM The output is AV
1260 AV=0
1270 TERM=1
1280 FOR I=0 TO R-1
1290 IF I > A THEN AV=AV+TERM*(1-P)**(N-I)*P**I
1300 TERM=TERM*(N-I)/(I+1)
1310 NEXT I
1320 AV=2*N*AV+N*(1-AV)
1330 RETURN
1340 REM
1350 REM This GOSUB calculates the AOQL
1360 REM using a quadratic approximation
1370 P0=0
1380 AV0=0
1390 P2=(A+1)/N/8
1400 P=P2
1410 GOSUB 990
1420 AV2=P2*PA
1430 P1=(A+1)/N/4
1440 P=P1
1450 GOSUB 990
1460 AV1=P1*PA
1470 REM Order according to the value of AV*
1480 IF AV0>=AV1 THEN GOTO 1550
1490 AV=AV0
1500 AV0=AV1
1510 AV1=AV
1520 P=P0
1530 P0=P1
1540 P1=P
1550 IF AV0>=AV2 THEN GOTO 1620
1560 AV=AV0
1570 AV0=AV2
1580 AV2=AV
1590 P=P0
1600 P0=P2
1610 P2=P
1620 IF AV1>=AV2 THEN GOTO 1690
1630 AV=AV1
1640 AV1=AV2
1650 AV2=AV
1660 P=P1
1670 P1=P2
1680 P2=P
```

152 Quality assurance

```
1690 REM if difference of max and min < .001 then return
1700 IF AVO-AV2 <= .001 THEN GOTO 1810
1710 DO=(AV1-AV2)*PO
1720 D1=(AV2-AVO)*P1
1730 D2=(AVO-AV1)*P2
1740 NM=PO*DO+P1*D1+P2*D2
1750 DEN=DO+D1+D2
1760 P2=NM/DEN/2
1770 P=P2
1780 GOSUB 990
1790 AV2=P2*PA
1800 GOTO 1470
1810 PRINT
1820 PRINT '                                        The AOQL is';AVD
1830 PRINT
1840 RETURN
1850 REM
1860 REM this section calculates the maximum
1870 REM value of the average number inspected
1880 PO=0
1890 AVO=0
1900 P1=(A+1)/N
1910 P=P1
1920 GOSUB 1230
1930 AV1=AV
1940 P2=2*(A+1)/N
1950 P=P2
1960 GOSUB 1230
1970 AV2=AV
1980 REM Order according to value of AV*
1990 IF AVO>=AV1 THEN GOTO 2060
2000 AV=AVO
2010 AVO=AV1
2020 AV1=AV
2030 P=PO
2040 PO=P1
2050 P1=P
2060 IF AVO>=AV2 THEN GOTO 2130
2070 AV=AVO
2080 AVO=AV2
2090 AV2=AV
2100 P=PO
2110 PO=P2
2120 P2=P
2130 IF AV1>=AV2 THEN GOTO 2200
2140 AV=AV1
2150 AV1=AV2
2160 AV2=AV
2170 P=P1
2180 P1=P2
2190 P2=P
2200 REM If difference is less than .01 then return
2210 IF AVO-AV2 <= .01 THEN GOTO 2320
2220 DO=(AV1-AV2)*PO
2230 D1=(AV2-AVO)*P1
2240 D2=(AVO-AV1)*P2
2250 NM=PO*DO+P1*D1+P2*D2
2260 DEN=DO+D1+D2
2270 P2=NM/DEN/2
2280 P=P2
2290 GOSUB 1230
2300 AV2=AV
```

Multiple sampling schemes   153

```
2310 GOTO 1980
2320 PRINT
2330 PRINT '      The maximum value of the AVERAGE number inspected is';AV2
2340 PRINT '      And is achieved when the proportion of defectives is';P2
2350 PRINT
2360 RETURN
2370 REM
2380 REM
2390 REM this GOSUB calculates P as a function of P(A)
2400 REM using linear interpolation
2410 PRINT '                          Input your value of P(A)';
2420 INPUT PB
2430 IF 0<=PB AND PB<=1 THEN GOTO 2460
2440 PRINT ' Iput error! Try again.'
2450 GOTO 2410
2460 D=.1
2470 P=0
2480 PAO=1
2490 P=P+D
2500 GOSUB 990
2510 IF PA <= PB THEN GOTO 2540
2520 PAO=PA
2530 GOTO 2490
2540 REM The value required lies between P and P-D
2550 REM Linear interpolation is used to find it
2560 PL=P-D
2570 PU=P
2580 PAL=PAO
2590 PAU=PA
2600 PN=PL+(PU-PL)*(PAO-PB)/(PAO-PA)
2610 P=PN
2620 GOSUB 990
2630 IF ABS(PA-PB) < .01 THEN GOTO 2710
2640 IF PA > PB THEN GOTO 2680
2650 PU=PN
2660 PAU=PA
2670 GOTO 2600
2680 PL=PN
2690 PAL=PA
2700 GOTO 2600
2710 PRINT
2720 PRINT '                          The value of P is';PN
2730 RETURN
2740 END
```

Program 7.2 Analysing sampling plans

Index

Accept, 141, 142
Ageing stress, 14
And gate, 79, 80, 81, 83
AOQ, 145, 146
AOQL, 144, 146
Assign, 2
Availability, 14, 44, 56
Average outgoing quality, 144
Average outgoing quality limit, 144

Basic event, 79, 80, 81, 82
Batch, 141, 142
Bayes, 106
Bayes' theorem, 106, 107, 108, 136
Bayesian analysis, 94, 106, 117, 118, 125, 140
Bayesian limit, 112, 114, 116, 137
Beta, 110, 115, 116, 117
Binomial, 96, 115, 118, 134, 142
Bottom up, 68
Burn in, 23

Censored, 35
CFR, 24, 26, 28, 34, 36, 41
Characteristic life, 26
Classical analysis, 126
Cold redundancy, 40
Combination event, 80
Confidence limit, 103, 134, 135, 136
Conjugate, 112, 113, 115
Constant failure rate, 24, 25
Counting variable, 6
Criticality, 68, 72
Cumulative frequency, 20
Customer's risk, 142
Cut set, 80, 81, 82

Decreasing failure rate, 24
Defect rate, 67
Degraded state, 47
Degradation, 47

Degraded system, 72
Design, 14, 15
Development, 14, 15
Development cycle, 14, 15
Derating, 68
DFR, 24, 28
DIM statement, 8
Down time, 43
Duane, 88, 91

Environment, 14
Ergonomics, 13, 84
Erlang, 98, 135
Event, 79, 80
Exponential, 28, 98, 118

Failure mode, 61, 68, 72
Failure mode ratio, 69, 72
Failure rate, 22, 23, 72
Feasibility study, 16
FOR NEXT loop, 6
Function, 13

Gamma, 109, 110, 112, 114, 117, 118, 137
Gate, 79
Gaussian elimination, 51

Hazard rate, 22
Hot redundancy, 40

IFR, 24, 28
Increasing failure rate, 24
INPUT command, 3
Input event, 79
Inspection by attributes, 141

k-out-of-n, 40

Least squares, 88, 127
Likelihood, 108, 111, 113, 115, 137
Logic gates, 80

156   Index

Log-log paper, 88
Log normal, 29
Lot, 141
Lower quality level, 142
LQL, 142, 143, 146

Maintainability, 43
Maintenance policy, 19
Manufacture, 14, 15
Manufacturing quality, 68
Marginal, 107, 108
Maximum likelihood, 108
Mean, 7
Mean, geometric, 44
Mean time to failure, 23
Mean time to repair, 44
Measure of effectiveness, 56
Min. cut, 81, 82, 84
Mission, 13, 72
MLE, 108, 110, 111, 117, 118, 137, 140
Mortality of humans, 24
MTTF, 23, 25, 36, 37, 51, 138
MTTR, 44
Multiprecision variables, 118

Null hypothesis, 95

OC, 142, 143, 145, 146
One-out-of-two-twice, 60
Operating characteristic, 142
Operational cycle, 14
Or gate, 79, 80, 81, 83
Output, 2
Output event, 79

Poisson, 96, 106
Poisson probability paper, 45, 115,
    118, 134, 137
Posterior, 108, 110, 111, 114, 116, 117
PRAT, 17
Probability density function, 22
Probability, laws of, 20
Primary basic event, 79
Prior, 107, 108, 111, 112, 114, 115,
    116, 117, 136
Producer's risk, 142
Product rule, 20

Quality assurance, 15, 17, 32

Quality control, 15, 17

Random failure, 87
Random sample, 141
Redundancy at component level, 60
Redundancy at system level, 61
Redundant system, 40
Reject, 141, 142
Reliability growth, 16
Reliability structure, 84
Reliability trials, 17
REM command, 3
Replacement, 45

Safety related failure, 84
Safety system, 58
Scale parameter, 26
Secondary basic event, 79
Series system, 39
Severity level, 68, 72, 73
Shape parameter, 26
Significance, 94
Significance level, 95, 103
Specification, 14
Specified task, 13
Standby redundancy, 40
String, 7
Summation rule, 20

Times to failure, 19, 20
Top event, 79
Transition matrix, 49, 53
Truncated normal, 29
Two-parameter model, 26
Type I error, 102
Type II error, 102

Upper quality level, 142
Up times, 43
UQL, 142, 143, 146
Use, 14
User requirement, 14

Variance, 7

Weakest link, 26
Weibull, 26, 28, 35, 126, 127, 128, 140
Working environment, 67